ICE Design and Construct

Conditions of Contract

Second Edition

CONDITIONS OF CONTRACT AND FORMS OF TENDER, AGREEMENT AND BOND FOR USE IN CONNECTION WITH WORKS OF CIVIL ENGINEERING CONSTRUCTION

The Institution of Civil Engineers
Association of Consulting Engineers
Civil Engineering Contractors Association

Published by Thomas Telford Publishing, Thomas Telford Limited, 1 Heron Quay, London E14 4JD on behalf of:

The Chief Executive and Secretary
The Institution of Civil Engineers
One Great George Street
London SW1P 3AA

The Chief Executive
Association of Consulting Engineers
Alliance House
12 Caxton Street
London SWIH 0QL

The Director
Civil Engineering Contractors Association
Construction House
56–64 Leonard Street
London EC2A 4JX

The first edition was approved by The Institution of Civil Engineers, the Association of Consulting Engineers and the Federation of Civil Engineering Contractors.
First edition, January 1992; reprinted without amendments 1997
Guidance Note March 1995 reference CCSJC/GN/March 1995

The following amendments were approved by The Institution of Civil Engineers, the Association of Consulting Engineers and the Civil Engineering Contractors Association
Reference ICE/D&C/Tax/February 1998
ICE/D&C/HGCR/March 1998
ICE/D&C/Pub/September 1998
ICE/D&C/CPF/February 1999

The second edition was approved by The Institution of Civil Engineers, the Association of Consulting Engineers and the Civil Engineering Contractors Association, July 2000.

The Institution of Civil Engineers, the Association of Consulting Engineers and the Civil Engineering Contractors Association have, as sponsoring authorities, approved this revised, second, edition of the document commonly known as the ICE Design and Construct Conditions of Contract, for all works of civil engineering construction. A permanent joint committee has prepared and will keep under review the use of the document and will consider any suggestions for amendment, which should be addressed to the Chief Executive and Secretary (CCSJC), The Institution of Civil Engineers, One Great George Street, London SWIP 3AA. Revision to the document will be made when such action seems warranted.

9 8 7 6 5 4 3 2 1

ISBN 0 7277 3023 1

A catalogue record for this book is available from the British Library

Printed and bound in Great Britain by
Selwood Printing Ltd., Burgess Hill, West Sussex

CONTENTS

Associated publications and publications referred to in the Contract

Guidance Notes to the ICE Design and Construct Conditions of Contract 2nd Edition

ICE Arbitration Procedure 1997
ICE Appendix (2001) to "The Scottish Arbitration Code 1999"
ICE Conciliation Procedure 1999
ICE Adjudication Procedure (1997)

Other ICE Conditions of Contract publications

ICE Conditions of Contract 7th Edition
Guidance Notes to the ICE Conditions of Contract 7th Edition
ICE Conditions of Contract Minor Works 3rd Edition (including Guidance Notes)

CONTENTS OF THE ICE DESIGN AND CONSTRUCT CONDITIONS OF CONTRACT

INDEX

INDEX	CLAUSE	PAGE

ICE Design and Construct Conditions of Contract

DEFINITIONS AND INTERPRETATION

Definitions **1** (1) In the Contract (as hereinafter defined) the following words and expressions shall have the meanings hereby assigned to them except where the context otherwise requires

(a) "Employer" means the person or persons firm company or other body named in the Appendix to the Form of Tender and includes the Employer's personal representatives successors and permitted assignees.

(b) "Contractor" means the person or persons firm or company to whom the Contract has been awarded by the Employer and includes the Contractor's personal representatives successors and permitted assigns.

(c) "Employer's Representative" means the person appointed by the Employer to act as such for the purposes of the Contract or any other person so appointed from time to time by the Employer and notified in writing as such to the Contractor.

(d) "Contractor's Representative" means the person appointed by the Contractor to act as such for the purposes of the Contract and notified in writing as such to the Employer or other person so appointed from time to time by the Contractor and similarly notified.

(e) "Employer's Requirements" means the requirements which are identified as such at the date of the award of the Contract. Any subsequent alterations thereto and which may describe the standards performance and/or objectives that are to be achieved by the Works or parts thereof are deemed to constitute an alteration under Clause 51.

(f) "Contractor's Submission" means the tender and all documents forming part of the Contractor's offer together with such modifications and additions thereto as may be agreed between the parties prior to the award of the Contract

(g) "Contract" means the Conditions of Contract the Employer's Requirements the Contractor's Submission and the written acceptance thereof together with such other documents as may be expressly agreed between the parties and the Form of Agreement (if completed).

(h) "Contract Price" means the sum to be ascertained and paid for the design construction and completion of the Works in accordance with the Contract.

(i) "Prime Cost Item" means an item in the Contract which contains (either wholly or in part) a sum referred to as Prime Cost (PC) which will be used for the supply of goods materials or services for the Works.

(j) "Provisional Sum" means any sum included and so designated in the Contract as a specific contingency for the carrying out of work or the supply of goods materials or services which may be used in whole in part or not at all in accordance with the specific requirements stated therefor.

(k) "Permanent Works" means the permanent works to be constructed and completed in accordance with the Contract.

(l) "Temporary Works" means all temporary works of every kind required in or about the construction and completion of the Works.

(m) "Works" means the Permanent Works and the Temporary Works and the design of both.

(n) "Commencement Date". As defined in Clause 41(1).

(o) "Certificate of Substantial Completion" means a certificate issued under Clause 48.

(p) "Defects Correction Period" means that period stated in the Appendix to the Form of Tender calculated from the date on which the Contractor becomes entitled to a Certificate of Substantial Completion for the Works or any Section or part thereof.

(q) "Defects Correction Certificate". As defined in Clause 61(1).

(r) "Section" means a part of the Works separately identified in the Appendix to the Form of Tender.

(s) "Site" means the lands and other places on under in or through which the Works are to be constructed and any other lands or places provided by the Employer for the purposes of the Contract together with such other places as may be designated in the Contract or subsequently agreed by the Employer's Representative as forming part of the Site.

(t) "Contractor's Equipment" means all appliances or things of whatever nature required in or about the construction and completion of the Works but does not include materials or other things intended to form or forming part of the Permanent Works.

Singular and plural

(2) Words importing the singular also include the plural and vice-versa where the context requires.

Headings and marginal notes

(3) The headings and marginal notes in the Conditions of Contract shall not be deemed to be part thereof or be taken into consideration in the interpretation or construction thereof or of the Contract.

Clause references

(4) All references herein to clauses are references to clauses numbered in the Conditions of Contract and not to those in any other document forming part of the Contract.

Cost

(5) The word "cost" when used in the Conditions of Contract means all expenditure including design costs properly incurred or to be incurred whether on or off the Site including overhead finance and other charges (including loss of interest) properly allocatable thereto but does not include any allowance for profit.

Communications in writing

(6) Communications which under the Contract are required to be "in writing" may be hand-written typewritten or printed and sent by hand post telex cable facsimile or other means resulting in a permanent record.

Consent and approval

(7) (a) The giving of any consent or approval by or on behalf of the Employer's Representative shall not in any way relieve the Contractor of any of his obligations under the Contract or of his duty to ensure the correctness accuracy or suitability of the matter or thing which is the subject of the consent or approval.

(b) Failure by the Employer's Representative or any of his assistants to disapprove or object to any matter or thing shall not prejudice his power subsequently to take action under the Contract in connection therewith.

(c) Failure by the Employer's Representative to object to any method of construction Temporary Works or use of Contractor's Equipment notified pursuant to Clause 6(4) or to reject any programme submitted pursuant to Clause 14 shall not in any way imply that the method of construction Temporary Works use of Contractor's Equipment or programme in question is feasible suitable or appropriate and the Employer's rights under the Contract shall in not way be impaired thereby.

EMPLOYER'S REPRESENTATIVE

Duties and authority of Employer's Representative

2 (1) (a) The Employer's Representative shall carry out the duties and may exercise the authority specified in or necessarily to be implied from the Contract.

(b) Except as expressly stated in the Contract the Employer's Representative shall have no authority to amend the Contract nor to relieve the Contractor of any of his obligations under the Contract.

Named individual

(2) Within 7 days of the award of the Contract and in any event before the Commencement Date the Employer shall notify to the Contractor in writing the name of the Employer's Representative. The Employer shall similarly notify the Contractor of any replacement of the Employer's Representative.

Delegation by the Employer's Representative

(3) The Employer's Representative may from time to time delegate to any person (including assistants appointed under sub-clause (4) of this Clause) any of the duties and authorities vested in him and he may at any time revoke such delegation.

Any such delegation

(a) shall be in writing and shall not take effect until such time as a copy thereof has been delivered to the Contractor or a representative appointed under Clause 15

(b) shall continue in force until such time as the Employer's Representative shall notify the Contractor or his representative in writing that the same has been revoked and

(c) shall not be given in respect of any decision to be taken or certificate to be issued under Clauses 12(6) 13 15(2)(b) 44 48 60(4) 61 65 or 66.

Assistants

(4) (a) The Employer's Representative may appoint any number of assistants for the carrying out of his duties under the Contract. He shall notify to the Contractor the names duties and scope of authority of such assistants.

(b) Assistants may be appointed specifically to watch the construction of the Works.

(c) Assistants shall have no authority to relieve the Contractor of any of his duties or obligations under the Contract nor except as provided under sub-clause (3) of this Clause

(i) to order any work involving delay or any extra payment by the Employer or

(ii) to make any alteration of or in the Works.

Instructions

(5) (a) Instructions given by the Employer's Representative or by any person exercising delegated duties and authorities under sub-clause (3) of this Clause shall be in writing. Provided that if for any reason it is considered necessary to give any such instruction orally the Contractor shall comply therewith.

(b) Any such oral instruction shall be confirmed in writing by or on behalf of the Employer's Representative as soon as is possible in the circumstances. Provided that if the Contractor confirms in writing any such oral instruction which is not contradicted in writing by or on behalf of the Employer's Representative forthwith it shall be deemed to be an instruction in writing by the Employer's Representative.

(c) Upon the written request of the Contractor the Employer's Representative or the person exercising delegated duties or authorities under sub-clause (3) of this Clause shall specify in writing under which of his duties and authorities any instruction is given.

Impartiality

(6) The Employer's Representative shall exercise his functions under Clauses 12 13 44 46 47(6) 48 52 53 60 61 65 and 66 or any matter relevant to safety impartially within the terms of the Contract having regard to all the circumstances.

ASSIGNMENT AND SUB-CONTRACTING

Assignment **3**

(1) Neither the Employer nor the Contractor shall assign the Contract or any part thereof or any benefit or interest therein or thereunder without the prior written consent of the other party which consent shall not unreasonably be withheld.

Rights of third parties

(2) Nothing in this Contract confers or purports to confer on any third party any benefit or any right pursuant to the Contracts (Rights of Third Parties) Act 1999 to enforce any term of the Contract.

Sub-contracting **4**

(1) The Contractor shall not sub-contract the whole of the construction of the Works without the prior written consent of the Employer.

(2) (a) The Contractor must obtain the Employer's consent prior to making any change in the Contractor's designer named in the Appendix to the Form of Tender which consent shall not unreasonably be withheld.

(b) Except where otherwise provided in the Appendix to the Form of Tender the Contractor may sub-contract any part of the Works or their design. The extent of the work to be sub-contracted and the name and address of the sub-contractor must be notified in writing to the Employer's Representative as soon as practicable and in any event not later than 14 days prior to the sub-contractor's entry on to the Site or in the case of design on appointment.

Provided that if no later than 7 days after receipt of such notification the Employer's Representative for good reason objects to the Contractor in writing the sub-contractor so notified shall not be employed on or in connection with the Works. Such objection must be accompanied by reasons in writing.

(3) The employment of labour-only sub-contractors does not require notification to the Employer's Representative under sub-clause (2)(b) of this Clause.

(4) The Contractor shall be and remain liable under the Contract for all work sub-contracted and for acts defaults or neglects of any sub-contractor his agents servants or workpeople.

(5) The Employer's Representative may after due warning in writing require the Contractor to remove from the Works any sub-contractor who misconducts himself or is incompetent or negligent in the performance of his duties or fails to conform with any particular provisions with regard to safety which may be set out in the Contract or persists in any conduct which is prejudicial to safety or health and any such sub-contractor shall not be employed again upon the Works without the permission of the Employer's Representative.

DOCUMENTATION AND INFORMATION

Contract documents **5** (1) (a) The several documents forming the Contract are to be taken as mutually explanatory of one another.

(b) If in the light of the several documents forming the Contract there remain ambiguities or discrepancies between the Employer's Requirements and the Contractor's Submission the Employer's Requirements shall prevail.

(c) (i) Any ambiguities or discrepancies within the Employer's Requirements shall be explained and adjusted by the Employer's Representative who shall thereupon issue to the Contractor appropriate instructions in writing.

(ii) Should such instructions involve the Contractor in delay or disrupt his arrangements or methods of construction so as to cause him to incur cost beyond that reasonably to have been foreseen by an experienced contractor at the time of the award of the Contract then the Employer's Representative shall take such delay into account in determining any extension of time to which the Contractor is entitled under Clause 44 and the Contractor shall subject to Clause 53 be paid in accordance with Clause 60 the amount of such cost as may be reasonable. Profit shall be added thereto in respect of the design and construction of any additional permanent or temporary work.

(d) Any ambiguities or discrepancies within the Contractor's Submission shall be resolved at the Contractor's expense.

Supply of contract documents

(2) Upon the award of the Contract the Employer shall assemble two complete copies of the Contract of which one copy shall be supplied to the Contractor free of charge.

Design information

6 (1) (a) The Employer's Representative shall provide to the Contractor such design criteria relevant to the Employer's Requirements as may be necessary to enable the Contractor to comply with this Clause.

Further information

(b) The Contractor shall give the Employer's Representative notice in writing of any further information from the Employer that is required for the design and/or construction of the Works and to which the Contractor is entitled under the Contract.

(c) Should

(i) the Employer's Representative fail to provide any such design criteria or further information within a reasonable period following notice or

(ii) the criteria supplied pursuant to sub-clause (1)(a) of this Clause impose limitations which could not reasonably have been foreseen by an experienced contractor before the award of the Contract

the Employer's Representative shall take that failure or limitation into account in determining any extension of time to which the Contractor is entitled under Clause 44 and the Contractor shall subject to Clause 53 be paid in accordance with Clause 60 any reasonable cost which may arise therefrom.

Designs and drawings

(2) (a) The Contractor shall except as may otherwise be provided in the Contract submit to the Employer's Representative such designs and drawings as are necessary to show the general arrangement of the Works and that the Works will comply with the Employer's Requirements.

Construction to such designs and drawings shall not commence until the Employer's Representative has consented thereto.

(b) If in the opinion of the Employer's Representative any such design or drawing does not comply with the Employer's Requirements or with any other provision of the Contract he shall so inform the Contractor in writing giving his reasons and may withhold his consent thereto until the Contractor has re-submitted the design or drawing with appropriate modifications.

(c) The Contractor shall notify the Employer's Representative if he later wishes to modify any design or drawing to which consent has already been given and shall submit the modified design or drawing for further consent.

(d) Should the Employer's Representative fail within a reasonable period following the submission or re-submission of any design or drawing under this Clause to notify the Contractor either that he consents thereto or that he is withholding his consent he shall take such failure into account in determining any extension of time to which the Contractor is entitled under Clause 44 and the Contractor shall subject to Clause 53 be paid in accordance with Clause 60 any reasonable cost which may arise from such failure.

Supply of other information

(3) The Contractor shall supply to the Employer's Representative two copies of such other designs drawings specifications documents and information as the Employer's Representative may reasonably require.

Methods of construction

(4) If requested by the Employer's Representative the Contractor shall submit at such times and in such further detail as the Employer's Representative may reasonably require information pertaining to the methods of construction (including Temporary Works and the use of Contractor's Equipment) which the Contractor proposes to adopt or use.

Where appropriate such information shall include calculations of stresses strains and deflections that will arise in the Permanent Works or any parts thereof during construction so as to demonstrate to the Employer's Representative that if these methods are adhered to the Works can be constructed and completed in accordance with the Contract and without detriment to the Permanent Works when completed.

Copyright

7 (1) (a) Copyright of all drawings specifications and other documents made by or on behalf of the Employer shall as between the parties remain in the Employer. The Contractor may obtain or make at his own expense any further copies required by him for the purposes of the Contract.

(b) Copyright of all drawings specifications and other documents made by or on behalf of the Contractor shall as between the parties remain in the Contractor. The Employer may obtain or make at his own expense any further copies required by him for the purposes of the Contract and after completion of the Works for the operation maintenance dismantling reassembly repair alteration and adjustment of the Permanent Works.

Documents to be available on Site

(2) One copy of the Contract and such further documents as may be consented to under Clause 6(2) and all working drawings shall be kept by the Contractor on Site and the same shall be available at all reasonable times for inspection and use by the Employer's Representative his assistants and any other person authorised by the Employer's Representative in writing.

GENERAL OBLIGATIONS

Contractor's general obligations

8 (1) The Contractor shall subject to the provisions of the Contract and save in so far as it is legally or physically impossible

(a) design construct and complete the Works and

(b) provide all design services labour materials Contractor's Equipment Temporary Works transport to and from and in or about the Site and everything whether of a temporary or permanent nature required in and for such design construction and completion so far as the necessity for providing the same is specified in or reasonably to be inferred from the Contract.

Contractor's design responsibility

(2) (a) In carrying out all his design obligations under the Contract including those arising under sub-clause (2)(b) of this Clause (and including the selection of materials and plant to the extent that these are not specified in the Employer's Requirements) the Contractor shall exercise all reasonable skill care and diligence.

(b) Where any part of the Works has been designed by or on behalf of the Employer and that design has been identified in the Employer's Requirements that the Contractor shall to that extent check the design and accept the responsibility therefor having first obtained the approval of the Employer's Representative for any modifications thereto which the Contractor considers to be necessary.

Quality assurance

(3) To the extent required by the Contract the Contractor shall institute a quality assurance plan. The Contractor's quality plan shall be submitted to the Employer's Representative for his consent before each design and each construction stage is commenced.

Should the Employer's Representative fail within a reasonable period following the Contractor's submission or re-submission of the quality plan and procedures to notify the Contractor either that he consents to them or that he is withholding his consent he shall take such failure into account in determining any extension of time to which the Contractor is entitled under Clause 44 and the Contractor shall subject to Clause 53 be paid in accordance with Clause 60 any reasonable cost which may arise from such failure.

Compliance with such quality plan and procedures shall not relieve the Contractor from any of his other duties obligations or liabilities under the Contract.

Statutory checks

(4) Where any Act of Parliament Regulation or By-Law requires that a separate check of the design or a test shall be carried out prior to the construction or loading of any permanent and temporary works the Contractor shall arrange and pay for such check or test.

Contractor responsible for safety

(5) The Contractor shall take responsibility for the safety of the design and for the adequacy stability and safety of all site operations and methods of construction.

Form of agreement

9 The Contractor shall if called upon so to do enter into and execute an agreement to be prepared at the cost of the Employer in the form annexed to these Conditions.

Performance security

10 (1) If the Contract requires the Contractor to provide security for the proper performance of the Contract he shall obtain and provide to the Employer within 28 days of the award of the Contract such security in a sum not exceeding 10% of the estimated total value of the Works at the date of the award of the Contract. The security shall be provided by a body approved by the Employer and be in the Form of Bond annexed to these Conditions. The Contractor shall pay the cost of the security unless the Contract otherwise provides.

Dispute resolution upon security

(2) For the purposes of the dispute resolution provisions in such security

(a) the Employer shall be deemed to be a party to the security for the purpose of doing everything necessary to give effect to such provisions and

(b) any agreement decision award or other determination touching or concerning the relevant date for the discharge of the security shall be wholly without prejudice to the resolution or determination of any dispute between the Employer and the Contractor under Clause 66.

Provision and interpretation of information

11 (1) As between the Employer and the Contractor and without prejudice to sub-clause (2) of this Clause information on

(a) the nature of the ground and subsoil and hydrological conditions and

(b) pipes and cables in on or over the ground

obtained by or on behalf of the Employer from investigations undertaken relevant to the Works shall only be taken into account to the extent that it was made available to the Contractor before the Contractor's Submission.

The Contractor shall be responsible for the interpretation of all such information for the purposes of the Works.

Inspection of Site

(2) The Contractor shall be deemed to have inspected and examined the Site and its surroundings and information available in connection therewith and to have satisfied himself so far as is practicable and reasonable before the award of the Contract as to

(a) the form and nature thereof including the ground and subsoil and hydrological conditions and

(b) the extent and nature of work and materials necessary for constructing and completing the Works and

(c) the means of communication with and access to the Site and the accommodation he may require

and in general to have obtained for himself all necessary information as to risks contingencies and all other circumstances which may influence or affect his submission.

Basis and sufficiency of tender

(3) The Contractor shall be deemed to have

(a) submitted his tender on the basis of sub-clause (1) and (2) of this Clause and

(b) satisfied himself before the Contract is awarded as to the correctness and sufficiency of the rates and/or prices stated therein which shall(unless otherwise provided in the Contract) cover all his obligations under the Contract.

Adverse physical conditions and artificial obstructions

12 (1) If during the carrying out of the Works the Contractor encounters physical conditions (other than weather conditions or conditions due to weather conditions) or artificial obstructions which conditions or obstructions could not in his opinion reasonably have been foreseen by an experienced contractor the Contractor shall as early as practicable give written notice thereof to the Employer's Representative.

(2) If in addition the Contractor intends to make any claim for additional payment or extension of time arising from any such condition or obstruction he shall at the same time or as soon thereafter as may be reasonable inform the Employer's Representative in writing pursuant to Clause 53 and/or Clause 44(1) as may be appropriate specifying the condition or obstruction to which the claim relates.

Measures being taken

(3) When giving notification in accordance with sub-clauses (1) and/or (2) of this Clause or as soon as practicable thereafter the Contractor shall give details of any anticipated effects of the condition or obstruction the measures he has taken is taking or is proposing to take their estimated cost and extent of the anticipated delay in or interference with the carrying out of the Works. Such details shall include where appropriate consideration of alternative design measures and/or methods of procedure with comparative estimates of costs and delays.

Action by Employer's Representative

(4) Following receipt of any notification under sub-clauses (1) (2) or (3) of this Clause the Employer's Representative may if he thinks fit among other things

(a) give written consent to any measures with or without modification

(b) require the Contractor to investigate and report upon the practicality cost and timing of alternative measures which may be available

(c) order a suspension under Clause 40 or issue an order to alter the Employer's Requirements under Clause 51.

Conditions reasonably foreseeable

(5) If the Employer's Representative decides that any such condition or obstruction could in whole or in part have been reasonably foreseen by an experienced contractor he shall so inform the Contractor in writing as soon as he has reached that decision but the value of any alteration previously ordered by him pursuant to sub-clause (4)(b) of this Clause shall be ascertained in accordance with Clause 52 and paid for in accordance with Clause 60.

Delay and extra cost

(6) Where an extension of time or additional payment is claimed pursuant to sub-clause (2) of this Clause the Employer's Representative shall if in his opinion such conditions or obstructions could not reasonably have been foreseen by an experienced contractor

(a) determine any delay which the Contractor has suffered and

(b) determine the amount of any costs which may reasonably have been incurred by the Contractor (together with a reasonable percentage addition thereto in respect of profit)

by reason of such conditions or obstructions and shall notify the Contractor accordingly with a copy to the Employer.

Any delay so determined shall forthwith be considered under Clause 44(3) for an appropriate extension of time and the Contractor shall subject to Clause 53 be paid in accordance with Clause 60 the amount of any costs so determined.

Instructions of the Employer's Representative

13 (1) If in the opinion of the Employer's Representative the Contractor's conduct of the Works (including but not confined to the Temporary Works and/or methods of construction) does not comply with the requirements of the Contract he may after first consulting with the Contractor issue instructions in writing on any matter connected therewith (whether mentioned in the Contract or not). Such instructions shall be accompanied by reasons stating in what respects the Contractor's conduct fails to comply.

Provided that if action is urgently necessary the Employer's Representative may issue such instructions without consultation.

(2) The Contractor shall take such steps as may be necessary to comply with such instructions. Provided that the Contractor shall take instructions under this Clause only from the Employer's Representative.

(3) If in pursuance of this Clause the Employer's Representative issues instructions which involve the Contractor in delay or disrupt his arrangements methods of construction and/or Temporary Works so as to cause him to incur cost beyond that reasonably to have been foreseen by an experienced contractor at the time of tender then the Employer's Representative shall take such delay into account in determining any extension of time to which the Contractor is entitled under Clause 44 and the Contractor shall subject to Clause 53 be paid in accordance with Clause 60 the amount of such cost as may be reasonable except to the extent that such delay and extra cost result from the Contractor's default. Profit shall be added thereto in respect of any additional permanent or temporary work. If such instructions require or necessarily imply any alteration to the Employer's Requirements for any part of the Works the same shall be deemed to have been ordered pursuant to Clause 51.

Programme to be furnished

14 (1) (a) Within 21 days after the award of the Contract the Contractor shall submit to the Employer's Representative a programme showing the order in which he proposes to carry out the Works having regard to the provisions of Clause 42(1).

(b) At the same time the Contractor shall provide in writing for the information of the Employer's Representative a general description of the arrangements and methods of construction which the Contractor proposes to adopt for the carrying out of the Works.

(c) Should the Employer's Representative reject any programme under sub-clause (2) of this Clause the Contractor shall within 14 days of such rejection submit a revised programme.

Action by Employer's Representative

(2) The Employer's Representative may within 14 days after receipt of the Contractor's programme reject it in writing with reasons or request the Contractor to supply further information to clarify or substantiate the programme or to satisfy the Employer's Representative as to its reasonableness having regard to the Employer's Requirements and the Contractors other obligations under the Contract.

Further information

(3) The Contractor shall within 14 days after receiving from the Employer's Representative any request under sub-clause (2) of this Clause or within such further period as the Employer's Representative may allow provide the further information requested failing which the relevant programme shall be deemed to be rejected.

Upon receipt of such further information the Employer's Representative may within a further 14 days reject the programme in accordance with sub-clause (2) of this Clause.

Revision of programme

(4) Should it appear to the Employer's Representative at any time that the actual progress of the work does not conform with the programme referred to in sub-clause (1) and not rejected in accordance with sub-clause (2) or (3) of this Clause the Employer's Representative shall be entitled to require the Contractor to produce a revised programme showing such modification to that programme as may be necessary to ensure completion of the Works or any Section within the time for completion as defined in Clause 43 or extended time granted pursuant to Clause 44.

In such event the Contractor shall submit his revised programme within 14 days or within such further period as the Employer's Representative may allow. Thereafter the provisions of sub-clauses (2) and (3) of this Clause shall apply.

Contractor's superintendence

15 (1) The Contractor shall provide all necessary superintendence during the carrying out of the Works and for as long thereafter as the Employer's Representative may reasonably consider necessary.

Such superintendence shall be given by sufficient persons having adequate knowledge of the operations to be carried out (including the methods and techniques required the hazards likely to be encountered and methods of preventing accidents) for the satisfactory and safe carrying out of the Works.

Contractor's Representative

(2) (a) The Contractor's Representative shall exercise overall superintendence of the Works on behalf of the Contractor. He shall receive on behalf of the Contractor all consents approvals orders instructions and information given by the Employer's Representative.

The Contractor's Representative shall have the authority to act and to commit the Contractor as if he were the Contractor for the purposes of the Contract.

(b) The Contractor's Representative may delegate any of his responsibilities to a nominated deputy with the prior written agreement of the Employer's Representative. Information and decisions from any such nominated deputy shall be as if from the Contractor's Representative.

Contractor's agent

(3) The Contractor's Representative or a competent and authorised agent approved in writing by the Employer's Representative (which approval may at any time be withdrawn) shall superintend the construction of the Permanent and Temporary Works.

From the time that the Contractor commences work on Site the authorised agent

(a) shall give his whole time to such superintendence

(b) shall receive on behalf of the Contractor any consents approvals orders instructions and information from any of the assistants of the Employer's Representative and

(c) may also receive on behalf of the Contractor's Representative consents approvals orders instructions and information from the Employer's Representative.

Removal of Contractor's Employees

16 (1) The Contractor shall employ or cause to be employed in connection with the Works and in the superintendence thereof only persons who are careful skilled and experienced in their several professions trades and callings.

(2) The Employer's Representative shall be at liberty to object to and require the Contractor to remove or cause to be removed from the Works any person employed thereon who in the opinion of the Employer's Representative misconducts himself or is incompetent or negligent in the performance of his duties or fails to conform with any particular provisions with regard to safety which may be set out in the Contract or persists in any conduct which is prejudicial to safety or health and such persons shall not be employed again upon the Works without the permission of the Employer's Representative.

17 Not used.

Boreholes and exploratory excavation

18 (1) If during the performance of the Works the Contractor considers it necessary or desirable to make boreholes or to carry out exploratory excavations or investigations of the ground he shall apply to the Employer's Representative for permission so to do giving his reasons and details of his proposed methods. Such permission shall not unreasonably be withheld.

The Contractor shall comply with any conditions imposed by the Employer's Representative in relation thereto and shall furnish the Employer's Representative with copies of all information records and test results arising therefrom and of any expert opinion as may be provided in connection therewith.

(2) The cost of making such boreholes and carrying out such investigations and of all other matters connected therewith including making good thereafter to the satisfaction of the Employer's Representative shall be borne by the Contractor. Provided that if in the opinion of the Employer's Representative the boreholes excavations and investigations are a necessary consequence of

(a) a situation arising under Clause 12 they shall be paid for in accordance with that Clause or

(b) an alteration ordered under Clause 51 they shall for the purposes of payment be treated as part of that alteration and priced in accordance with Clause 52.

(3) If during the performance of the Works the Employer's Representative requires the Contractor to make boreholes or to carry out exploratory excavations or investigations of the ground such requirements shall be ordered in writing and shall be deemed to be an alteration ordered under Clause 51 unless a Provisional Sum or Prime Cost Item in respect of such anticipated work is included in the Contract.

Safety and security

19 (1) The Contractor shall throughout the progress of the Works have full regard for the safety of all persons entitled to be upon the Site and shall keep the Site (so far as the same is under his control) and the Works (so far as the same are not completed or occupied by the Employer) in an orderly state appropriate to the avoidance of danger to such persons and shall among other things in connection with the Works provide and maintain at his own cost all lights guards fencing warning signs and watching when and where necessary or required by the Employer's Representative or any of his assistants or by any competent statutory or other authority for the protection of the Works or for the safety and convenience of the public or others.

Employer's responsibilities

(2) If the Employer carries out work on the Site with his own workpeople he shall in respect of such work

(a) have full regard for the safety of all persons entitled to be on the Site and

(b) keep the Site in an orderly state appropriate to the avoidance of danger to such persons.

If the Employer employs other contractors on the Site he shall require them to have the same regard for safety and avoidance of danger.

Care of the Works **20** (1) (a) The Contractor shall save as in paragraph (b) hereof and subject to sub-clause (2) of this Clause take full responsibility for the care of the Works and for materials plant and equipment for incorporation therein from the Commencement Date until the date of issue of a Certificate of Substantial Completion for the whole of the Works when the responsibility for the said care shall pass to the Employer.

(b) If the Employer's Representative issues a Certificate of Substantial Completion for any Section or part of the Permanent Works the Contractor shall cease to be responsible for the care of that Section or part from the date of issue of that Certificate of Substantial Completion when the responsibility for the care of that Section or part shall pass to the Employer. Provided always that the Contractor shall remain responsible for any damage to such completed work caused by or as a result of his other activities on the Site.

(c) The Contractor shall take full responsibility for the care of any outstanding work and materials plant and equipment for incorporation therein which he undertakes to finish during the Defects Correction Period until such outstanding work has been completed.

Excepted risks (2) The Excepted Risks for which the Contractor is not liable are loss and damage to the extent that it is due to

(a) the use or occupation by the Employer his agents servants or other contractors (not being employed by the Contractor) of any part of the Permanent Works

(b) any fault defect error or omission in the design of the Works for which the Contractor is not responsible under the Contract

(c) riot war invasion act of foreign enemies or hostilities (whether war be declared or not).

(d) civil war rebellion revolution insurrection or military or usurped power

(e) ionizing radiations or contamination by radioactivity from any nuclear fuel or from any nuclear waste from the combustion of nuclear fuel radioactive toxic explosive or other hazardous properties of any explosive nuclear assembly or nuclear component thereof and

(f) pressure waves caused by aircraft or other aerial devices travelling at sonic or supersonic speeds.

Rectification of loss or damage (3) (a) In the event of any loss or damage to

(i) the Works or any Section or part thereof or

(ii) materials plant or equipment for incorporation therein

while the Contractor is responsible for the care thereof (except as provided in sub-clause (2) of this Clause) the Contractor shall at his own cost rectify such loss or damage so that the Permanent Works conform in every respect with the provisions of the Contract. The Contractor shall also be liable for any loss or damage to the Works occasioned by him in the course of any operations carried out by him for the purpose of complying with his obligations under Clauses 49 and 50.

(b) Should any loss or damage arise from any of the Excepted Risks defined in sub-clause (2) of this Clause the Contractor shall if and to the extent required by the Employer' Representative rectify the loss or damage at the expense of the Employer.

(c) In the event of loss or damage arising from an Excepted Risk and a risk for which the Contractor is responsible under sub-clause (1)(a) of this Clause then the cost of rectification shall be apportioned accordingly.

Insurance of Permanent and Temporary Works etc

21 (1) The Contractor shall without limiting his or the Employer's obligations and responsibilities under Clause 20 insure in the joint names of the Contractor and the Employer the Permanent and Temporary Works together with materials plant and equipment for incorporation therein to the full replacement cost plus an additional 10% to cover any additional costs that may arise incidental to the rectification of any loss or damage including professional fees cost of demolition and removal of debris.

Extent of cover

(2) (a) The insurance required under sub-clause (1) of this Clause shall cover the Employer and the Contractor against all loss or damage from whatsoever cause arising other than the Excepted Risks defined in Clause 20(2) from the Commencement Date until the date of issue of the relevant Certificate of Substantial Completion.

(b) The insurance shall extend to cover any loss or damage arising during the Defects Correction Period from a cause occurring prior to the issue of any Certificate of Substantial Completion and any loss or damage occasioned by the Contractor in the course of any operation carried out by him for the purpose of complying with his obligations under Clauses 49 and 50.

(c) Nothing in this Clause shall render the Contractor liable to insure against the necessity for the repair or reconstruction of any work constructed with materials or workmanship not in accordance with the requirements of the Contract unless the Contract otherwise requires.

(d) Any amounts not insured or not recovered from insurers whether as excesses carried under the policy or otherwise shall be borne by the Contractor or the Employer in accordance with their respective responsibilities under Clause 20.

Damage to persons or property

22 (1) The Contractor shall except if and so far as the Contract provides otherwise and subject to the exceptions set out in sub-clause (2) of this Clause indemnify and keep indemnified the Employer against all losses and claims in respect of

(a) death of or injury to any person or

(b) loss of or damage to any property (other than the Works)

which may arise out of or in consequence of the design and construction of the Works and the remedying of any defects therein and against all claims demands proceedings damages costs charges and expenses whatsoever in respect thereof or in relation thereto.

Exceptions

(2) The exceptions referred to in sub-clause (1) of this Clause which are the responsibility of the Employer are

(a) damage to crops being on the Site (save in so far as possession has not been given to the Contractor)

(b) the use or occupation of land provided by the Employer for the purposes of the Contract (including consequent losses of crops) or interference whether temporary or permanent with any right of way light air or water or other easement or quasi-easement which is the unavoidable result of the construction of the Works in accordance with the Contract

(c) the right of the Employer to construct the Works or any part thereof on over under in or through any land

(d) damage which is the unavoidable result of the construction of the Works in accordance with the Employer's Requirements including any design for which the Contractor is not responsible under the Contract

(e) death of or injury to persons or loss of or damage to property resulting from any act neglect or breach of statutory duty done or committed by the Employer his agents servants or other contractors (not being employed by the Contractor) or for or in respect of any claims demands proceedings damages costs charges and expenses in respect thereof or in relation thereto.

Indemnity by Employer

(3) The Employer shall subject to sub-clause (4) of this Clause indemnify the Contractor against all claims demands proceedings damages costs charges and expenses in respect of the matters referred to in the exceptions defined in sub-clause (2) of this Clause.

Shared responsibility

(4) (a) The Contractor's liability to indemnify the Employer under sub-clause (1) of this Clause shall be reduced in proportion to the extent that the act or neglect of the Employer his agents' servants or other contractors (not being employed by the Contractor) may have contributed to the said death injury loss or damage.

(b) The Employer's liability to indemnify the Contractor under sub-clause (3) of this Clause in respect of matters referred to in sub-clause (2)(e) of this Clause shall be reduced in proportion to the extent that the act or neglect of the Contractor or his sub-contractors servants or agents may have contributed to the said death injury loss or damage.

Third party insurance **23**

(1) The Contractor shall without limiting his or the Employer's obligations and responsibilities under Clause 22 insure in the joint names of the Contractor and the Employer against liabilities for death of or injury to any person (other than any operative or other person in the employment of the Contractor or any of his sub-contractors) or loss of or damage to any property (other than the Works) arising out of the performance of the Contract other than those liabilities arising out of the exceptions defined in Clause 22 (2) (a)(b)(c) and (d).

(2) The insurance policy shall include a cross liability clause such that the insurance shall apply to the Contractor and to the Employer as separate insured.

(3) Such insurance shall be for at least the amount stated in the Appendix to the Form of Tender.

Accident or injury to operatives etc

24 The Employer shall not be liable for or in respect of any damages or compensation payable at law in respect or in consequence of any accident or injury to any operative or other person in the employment of the Contractor or any of his sub-contractors save and except and to the extent that such accident or injury results from or is contributed to by any act or default of the Employer his agents or servants and the Contractor shall indemnify and keep indemnified the Employer against all such damages and compensation (save and except as aforesaid) and against all claims demands proceedings costs charges and expenses whatsoever in respect thereof or in relation thereto.

Evidence and terms of insurance

25 (1) The Contractor shall provide satisfactory evidence to the Employer prior to the Commencement Date that the insurances required under the Contract have been effected and shall if so required produce the insurance policies for inspection. The terms of all such insurances shall be subject to the approval of the Employer (which approval shall not unreasonably be withheld). The Contractor shall upon request produce to the Employer receipts for the payment of current insurance premiums.

Excesses

(2) Any excesses on the policies of insurance effected under Clauses 21 and 23 shall be as stated by the Contractor in the Appendix to the Form of Tender.

Remedy on Contractor's failure to insure

(3) If the Contractor fails upon request to produce to the Employer satisfactory evidence that there is in force any of the insurances required under the Contract then the Employer may effect and keep in force any such insurance and pay such premium or premiums as may be necessary for that purpose and from time to time deduct the amount so paid from any monies due or which may become due to the Contractor or recover the same as a debt due from the Contractor.

Compliance with policy conditions

(4) Both the Employer and the Contractor shall comply with all conditions laid down in the insurance policies. Should the Contractor or the Employer fail to comply with any condition imposed by the insurance policies effected pursuant to the Contract each shall indemnify the other against all losses and claims arising from such failure.

Giving of notices and payment of fees

26 (1) Except where otherwise provided in the Contract the Contractor shall give all notices and pay all fees required to be given or paid by any Act of Parliament or any Regulation or By-Law of any local or other statutory authority in relation to the design construction and completion of the Permanent Works and the Temporary Works and by the rules and regulations of all public bodies and companies whose property or rights are or may be affected in any way by the Works.

Repayment by Employer

(2) The Employer shall repay or allow to the Contractor all such sums as the Employer's Representative certifies to have been properly payable and paid by the Contractor in respect of such fees and also all rates and taxes paid by the Contractor in respect of the Site or any part thereof or anything constructed or erected thereon or any part thereof or any temporary structures situated elsewhere but used exclusively for the purpose of the Works or any structures used temporarily and exclusively for the purposes of the Works.

Contractor to conform with Statutes etc.

(3) The Contractor shall ascertain and conform in all respects with the provisions of any general or local Act of Parliament and the Regulations and By-Laws of any local or other statutory authority which may be applicable to the Works and with such rules and regulations of public bodies and companies as aforesaid and shall keep the Employer indemnified against all penalties and liability of every kind for breach of any Act Regulation or Bye-Law.

Provided always that

(a) the Contractor shall not be required to indemnify the Employer against the consequences of any such breach which is the unavoidable result of complying with the Employer's Requirements or instructions of the Employer's Representative

(b) if the Employer's Requirements are at any time found not to be in conformity with any such Act Regulation or Bye-Law the Employer's Representative shall issue such instructions including the ordering of an alteration under Clause 51 as may be necessary to ensure conformity with such Act Regulation or Bye-Law

(c) unless the Contract otherwise provides the Contractor shall not be responsible for obtaining any planning permission which may be necessary in respect of the Permanent Works or of any Temporary Works designed other than by or on behalf of the Contractor and the Employer shall be deemed to warrant that all such permissions have been or will in due time be obtained.

(4) If the Contractor incurs delay or extra cost arising from matters referred to in sub-clause (3)(b) or failure of the Employer to comply with sub-clause (3)(c) of this Clause the Employer's Representative shall take such delay into account in determining any extension of time to which the Contractor may be entitled under Clause 44 and the Contractor shall subject to Clause 53 be paid in accordance with Clause 60 the amount of such extra cost as may be reasonable except to the extent that such delay or extra cost result from the Contractor's default.

New Roads and Street Works Act 1991— Definitions

27 (1) (a) In this Clause "the Act" shall mean the New Roads and Street Works Act 1991 and any statutory modification or re-enactment thereof for the time being in force.

(b) For the purpose of obtaining any licence under the Act required for the Permanent Works the undertaker shall be the Employer who for the purposes of the Act will be the licensee.

Provided that where the license contains a prohibition against assignment which is notified to the Contractor then the Contractor shall give the Employer all notices required to be given by the undertaker and shall indemnify the Employer from and against all costs and charges which may arise from any failure by him so to do.

(c) All other expressions common to the Act and to this Clause shall have the same meaning as those assigned to them by the Act.

Licences

(2) (a) The Employer shall obtain any street works licence and any other consent licence or permission that may be required for the carrying out of the Permanent Works and shall supply the Contractor with copies thereof including details of any conditions or limitations imposed.

(b) Any condition or limitation in any licence obtained after the award of the Contract shall be deemed to be an instruction under Clause 13.

Notices

(3) The Contractor shall be responsible for giving to any relevant authority any required notice (or advance notice where prescribed) of his proposal to commence any work. A copy of each such notice shall be given to the Employer.

Delays attributable to alterations

(4) If any instruction pursuant to sub-clause (2)(b) of this Clause results in delay to the construction and completion of the Works because the Contractor needs to comply with sub-clause (3) of this Clause the Employer's Representative shall in addition to valuing the alteration under Clause 52 take such delay into account in determining any extension of time to which the Contractor is entitled under Clause 44 and the Contractor shall subject to Clause 53 be paid in accordance with Clause 60 such additional cost as the Employer's Representative shall consider to have been reasonably attributable to such delay.

Patent rights

28 (1) The Contractor shall indemnify and keep indemnified the Employer from and against all claims and proceedings for or on account of infringement of any patent right design trademark or name or other protected right in respect of

(a) the design construction and completion of the Works or any part thereof or

(b) any Contractor's Equipment used for or in connection with the Works or

(c) any materials plant or equipment for incorporation in the Works

and from and against all claims demands proceedings damages costs charges and expenses whatsoever in respect thereof or in relation thereto except as provided in sub-clause (2) of this Clause.

(2) In like manner the Employer shall indemnify and keep indemnified the Contractor

(a) where such infringement results from compliance with any design or specification provided other than by or on behalf of the Contractor or

(b) where the infringement is due to the use of the Works or any part thereof in association or in combination with any artifact not supplied by or on behalf of the Contractor or

(c) where the infringement is due to the use of the Works (whether by the Employer or some other person or legal entity having the lawful use thereof) other than for a purpose indicated by or reasonably to be inferred from the Employer's Requirements

from and against all claims and proceedings for or on account of such infringement and from and against all claims demands proceedings damages costs charges and expenses whatsoever in respect thereof or in relation thereto.

Royalties

(3) The Contractor shall pay all tonnage and other royalties rent and other payments or compensation (if any) for getting stone sand gravel clay or other materials required for the Works.

Interference with traffic and adjoining properties

29 (1) All operations necessary for the construction and completion of the Works shall so far as compliance with the requirements of the Contract permits be carried on so as not to interfere unnecessarily or improperly with

(a) the convenience of the public or

(b) the access to public or private roads footpaths or properties whether in the possession of the Employer or of any other person and with the use or occupation thereof.

The Contractor shall indemnify and keep indemnified the Employer in respect of all claims demands proceedings damages costs charges and expenses whatsoever arising out of or in relation to any such matters.

Noise disturbance and pollution

(2) All work shall be carried out without unreasonable noise disturbance or other pollution.

Indemnity by Contractor

(3) To the extent that noise disturbance or other pollution is not the unavoidable consequence of constructing and completing the Works or performing the Contract the Contractor shall indemnify the Employer from and against any liability for damages on that account and against all claims demands proceedings damages costs charges and expenses whatsoever in regard or in relation to such liability

Indemnity by Employer

(4) The Employer shall indemnify the Contractor from and against any liability for damages on account of noise disturbance or other pollution which is the unavoidable consequence of carrying out the Permanent and Temporary Works and from and against all claims demands proceedings damages costs charges and expenses whatsoever in regard or in relation to such liability.

Avoidance of damage to highways etc.

30 (1) The Contractor shall use every reasonable means to prevent any of the highways or bridges communicating with or on the routes to the Site from being subjected to extraordinary traffic within the meaning of the Highways Act 1980 or in Scotland the Roads (Scotland) Act 1984 or any statutory modification or re-enactment thereof by any traffic of the Contractor or any of his sub-contractors.

In particular the Contractor shall select routes and use vehicles and restrict and distribute loads so that any such extraordinary traffic as will inevitably arise from the moving of Contractor's Equipment and materials or manufactured or fabricated articles from and to the Site shall be limited as far as reasonably possible and so that no unnecessary damage or injury may be occasioned to such highways and bridges.

Transport of Contractor's Equipment

(2) Save in so far as the Contract otherwise provides the Contractor shall be responsible for and shall pay the cost of strengthening any bridges or altering or improving any highway communicating with the Site to facilitate the movement of Contractor's Equipment or Temporary Works or materials therefor required in the construction and completion of the Works.

The Contractor shall indemnify and keep indemnified the Employer against all claims for damage to any highway or bridge communicating with the Site caused by such movement including such claims as may be made by any competent authority directly against the Employer pursuant to any Act of Parliament or other Statutory Instrument and shall negotiate and pay all claims arising solely out of such damage.

Transport of materials

(3) (a) If notwithstanding sub-clause (1) of this Clause any damage occurs to any bridge or highway communicating with the Site arising from the transport of materials or manufactured or fabricated articles being part of the Permanent Works the Contractor shall notify the Employer's Representative as soon as he becomes aware of such damage or as soon as he receives any claim from the authority entitled to make such a claim whichever is the earlier.

(b) Where under any Act of Parliament or Statutory Instrument the haulier of such materials or manufactured or fabricated articles is required to indemnify the highway authority against damage the Employer shall not be liable for any costs charges or expenses in respect thereof or in relation thereto.

(c) In all cases under this sub-clause other than those covered in paragraph (b) hereof the Employer shall negotiate the settlement of and pay all sums due in respect of such claims and shall indemnify the Contractor in respect thereof and in respect of all claims demands proceedings damages costs charges and expenses in relation thereto.

Provided always that if and so far as any such claim or part thereof is in the opinion of the Employer's Representative due to any failure on the part of the Contractor to observe and perform his obligations under sub-clause (1) of this Clause then the amount certified by the Employer's Representative to be due to such failure shall be paid by the Contractor to the Employer or deducted from any sum due or which may become due to the Contractor.

Facilities for other Contractors

31 (1) The Contractor shall in accordance with the requirements of the Employer's Representative afford all reasonable facilities for any other contractors employed by the Employer and for the employees of the Employer and of any other properly authorised authorities or statutory bodies who may be employed in the execution on or near the Site of any work not in the Contract.

Delay and extra cost

(2) If compliance with sub-clause (1) of this Clause involves the Contractor in delay or cost beyond that reasonably to be foreseen by an experienced contractor at the time of tender then the Employer's Representative shall take such delay into account in determining any extension of time to which the Contractor is entitled under Clause 44 and the Contractor shall subject to Clause 53 be paid in accordance with Clause 60 the amount of such cost as may be reasonable. Profit shall be added thereto in respect of any additional permanent or temporary work.

Fossils etc.

32 All fossils coins articles of value or antiquity and structures or other remains or things of geological or archaeological interest discovered on the Site shall as between the Employer and the Contractor be deemed to be the absolute property of the Employer and the Contractor shall take reasonable precautions to prevent his employees or any other person from removing or damaging any such article or thing and shall immediately upon discovery thereof and before removal inform the Employer's Representative of such discovery and carry out at the expense of the Employer the Employer's Representative's orders as to the examination and disposal of the same.

Clearance of Site on completion

33 On completion of the Works the Contractor shall clear away and remove from the Site all Contractor's Equipment surplus material rubbish and Temporary Works of any kind and leave the whole of the Site and Permanent Works clean and in workmanlike condition to the satisfaction of the Employer's Representative.

34 Not Used.

Returns of labour and Contractor's Equipment

35 (1) If the Contract so provides the Contractor shall deliver to the Employer's Representative a return in such form and at such intervals as the Employer's Representative may prescribe showing in detail the numbers of the several classes of labour from time to time employed by the Contractor on the Site and such information in respect of Contractor's Equipment as the Employer's Representative may require. The Contractor shall require his sub-contractors to observe the provisions of this Clause.

(2) If the Contract contains no such provision the Contractor shall nevertheless and if so requested in writing by the Employer's Representative deliver a return in like manner in connection with any alteration claim or event.

MATERIALS AND WORKMANSHIP

Materials and workmanship **36** (1) The Works shall be designed constructed and completed in accordance with the Contract and where not expressly provided otherwise in the Contract in accordance with appropriate standards and standard codes of practice.

(2) All materials and workmanship shall be of the respective kinds described in the Contract or where not so described shall be appropriate in all the circumstances.

Checks and tests

(3) (a) Further to his obligations under Clause 8(3) the Contractor shall submit to the Employer's Representative for his approval proposals for checking the design and setting out of the Works and testing the materials and workmanship to ensure that the Contractor's obligations under the Contract are met.

b) The Contractor shall carry out the checks and tests approved under sub-clause (3)(a) of this Clause or elsewhere in the Contract and such further tests as the Employer's Representative may reasonably require.

Tests

(4) The Contractor shall provide such assistance and such instruments machines labour and materials as are normally required for examining measuring and testing any work and the quality weight or quantity of any materials used and shall supply samples of materials before incorporation in the Works for testing as may be required by the Employer's Representative.

Cost of samples

(5) All samples shall be supplied by the Contractor at his own cost if the supply thereof is clearly intended by or provided for in the Contract but if not then at the cost of the Employer.

Tests following alterations

(6) Whenever an alteration to the Employer's Requirements is ordered or consented to by the Employer's Representative the Contractor shall consider whether any tests would be affected by or be appropriate in relation thereto and shall so inform the Employer' Representative without delay. Any proposal for amended or additional tests shall be submitted as soon as possible.

Cost of tests

(7) Unless the Contract otherwise provides the cost of making any test shall be borne by the Contractor if such test is

(a) proposed by the Contractor under Clause 8(3) or sub-clause (3)(a) of this Clause or

(b) clearly intended by or provided for in the Contract.

If any test is carried out pursuant to sub-clause (3)(b) of this Clause the cost of such test shall be borne by the Contractor if the test shows the workmanship or materials not to be in accordance with the provisions of the Contract but otherwise by the Employer.

Access to Site **37** The Employer's Representative and any person authorised by him shall at all times have access to the Works and to the Site and to all workshops and places where work is being prepared or whence materials manufactured articles and machinery are being obtained for the Works and the Contractor shall afford every facility for and every assistance in obtaining such access or the right to such access.

Examination of work before covering up

38 (1) The Contractor shall afford full opportunity for the Employer's Representative to examine and measure any work which is about to be covered up or put out of view and to examine foundations before permanent work is placed thereon.

The Contractor shall give due notice to the Employer's Representative whenever any such work or foundations is or are ready or about to be ready for examination and the Employer's Representative shall without unreasonable delay attend for the purpose of examining and measuring such work or of examining such foundations unless he considers it unnecessary and advises the Contractor accordingly.

Uncovering and making openings

(2) The Contractor shall uncover any part or parts of the Works or make openings in or through the same as the Employer's Representative may from time to time direct and shall reinstate and make good such part or parts to the satisfaction of the Employer's Representative.

If any such part or parts have been covered up or put out of view after compliance with the requirements of sub-clause (1) of this Clause and are found to have been carried out in accordance with the Contract the cost of uncovering making openings in or through reinstating and making good the same shall be borne by the Employer but in any other case all such cost shall be borne by the Contractor.

Removal of unsatisfactory work and materials

39 (1) The Employer's Representative shall during the progress of the Works have power to instruct in writing the

(a) removal from the Site within such time or times specified in the instruction of any materials which in the opinion of the Employer's Representative are not in accordance with the Contract

(b) substitution with materials in accordance with the Contract and

(c) removal and proper replacement (notwithstanding any previous test thereof or interim payment therefor) of any work which in respect of

(i) materials or workmanship or

(ii) design by the Contractor or for which he is responsible

does not in the opinion of the Employer's Representative comply with the Contract.

Default of Contractor in compliance

(2) In case of default on the part of the Contractor in carrying out such instruction the Employer shall be entitled to employ and pay other persons to carry out the same and all costs consequent thereon or incidental thereto as determined by the Employer's Representative shall be recoverable from the Contractor by the Employer and may be deducted by the Employer from any monies due or to become due to him and the Employer's Representative shall notify the Contractor accordingly with a copy to the Employer.

Failure to disapprove

(3) Failure of the Employer's Representative or any person acting under him pursuant to Clause 2 to disapprove any work or materials shall not prejudice the power of the Employer's Representative or any such person subsequently to take action under this Clause.

Suspension of Works **40** (1) The Contractor shall on the written order of the Employer's Representative suspend the progress of the Works or any part thereof for such time or times and in such manner as the Employer's Representative may consider necessary and shall during such suspension properly protect and secure the work so far as is necessary in the opinion of the Employer's Representative. Except to the extent that such suspension is

(a) otherwise provided for in the Contract or

(b) necessary by reason of weather conditions or by some default on the part of the Contractor or

(c) necessary for the proper construction and completion or for the safety of the Works or any part thereof in as much as such necessity does not arise from any act or default of the Employer's Representative or the Employer or from any of the Excepted Risks defined in Clause 20(2).

then if compliance with the Employer's Representative's instructions under this clause involves the Contractor in delay or extra cost the Employer's Representative shall take such delay into account in determining any extension of time to which the Contractor is entitled under Clause 44 and the Contractor shall subject to Clause 53 be paid in accordance with Clause 60 the amount of such extra cost as may be reasonable. Profit shall be added thereto in respect of any additional permanent or temporary work.

Suspension lasting more than three months

(2) If the progress of the Works or any part thereof is suspended on the written order of the Employer's Representative and if permission to resume work is not given by the Employer's Representative within a period of 3 months from the date of suspension then the Contractor may unless such suspension is otherwise provided for in the Contract or continues to be necessary by reason of some default on the part of the Contractor serve a written notice on the Employer's Representative requiring permission within 28 days from the receipt of such notice to proceed with the Works or that part thereof in regard to which progress is suspended.

If within the said 28 days the Employer's Representative does not grant such permission the Contractor by a further written notice so served may elect to treat the suspension where it affects part only of the Works as an omission of such part under Clause 51 or where it affects the whole of the Works as an abandonment of the Contract by the Employer.

COMMENCEMENT AND DELAYS

Commencement Date **41** (1) The Commencement Date shall be

(a) the date specified in the Appendix to the Form of Tender or if no date is specified

(b) a date between 14 and 28 days of the award of the Contract to be notified to the Contractor by the Employer's Representative in writing or

(c) such other date as may be agreed between the parties.

Start of Works

(2) The Contractor shall start the Works on or as soon as is reasonably practicable after the Commencement Date. Thereafter the Contractor shall proceed therewith with due expedition and without delay in accordance with the Contract.

Possession of Site and access

42 (1) The Contract may prescribe

(a) the extent of portions of the Site of which the Contractor is to be given possession from time to time

(b) the order in which such portions of the Site shall be made available to the Contractor

(c) the availability and the nature of the access which is to be provided by the Employer and

(d) the order in which the Works shall be carried out.

(2) (a) Subject to sub-clause (1) of this Clause the Employer shall give to the Contractor on the Commencement Date possession of the whole of the Site together with such access thereto as may be necessary to enable the Contractor to commence and proceed with the carrying out of the Works.

(b) Thereafter the Employer shall during the course of the Works give to the Contractor such further access in accordance with the Contract as is necessary to enable the Contractor to carry out of the Works with due despatch.

Failure to give possession

(3) If the Contractor suffers delay and/or incurs extra cost from failure on the part of the Employer to give possession or access in accordance with the terms of this Clause the Employer's Representative shall take such delay into account in determining any extension of time to which the Contractor is entitled under Clause 44 and the Contractor shall subject to Clause 53 be paid in accordance with Clause 60 the amount of any extra cost to which he may be entitled. Profit shall be added thereto in respect of any additional Works.

Access and facilities provided by the Contractor

(4) The Contractor shall bear all costs and charges for any access required by him additional to that provided by the Employer. The Contractor shall also provide at his own cost any additional facilities outside the Site required by him for the purposes of the Works.

Time for completion

43 The whole of the Works and any Section required to be completed within a particular time as stated in the Appendix to the Form of Tender shall be substantially completed within the time so stated (or such extended time as may be allowed under Clause 44 or revised time agreed under Clause 46(3)) calculated from the Commencement Date.

Extension of time for completion

44 (1) Should the Contractor consider that

(a) any alteration ordered under Clause 51(1) or

(b) any cause of delay referred to in these Conditions or

(c) exceptional adverse weather conditions or

(d) any delay impediment prevention or default by the Employer or

(e) other special circumstances of any kind whatsoever which may occur be such as to entitle him to an extension of time for the substantial completion of the Works or any Section thereof he shall within 28 days after the cause of any delay has arisen or as soon thereafter as is reasonable deliver to the Employer's Representative full and detailed particulars in justification of the period of extension claimed in order that the claim may be investigated at the time.

Assessment of delay

(2) (a) The Employer's Representative shall upon receipt of such particulars consider all the circumstances known to him at that time and make an assessment of the delay (if any) that has been suffered by the Contractor as a result of the alleged cause and shall so notify the Contractor in writing with a copy to the Employer.

(b) The Employer's Representative may in the absence of any claim make an assessment of the delay that he considers has been suffered by the Contractor as a result of any of the circumstances listed in sub-clause (1) of this Clause and shall so notify the Contractor in writing.

Interim grant of extension of time

(3) Should the Employer's Representative consider that the delay suffered fairly entitles the Contractor to an extension of time for the substantial completion of the Works or any Section thereof such interim extension shall be granted forthwith and be notified to the Contractor in writing.

If the Contractor has made a claim for an extension of time but the Employer's Representative does not consider that the Contractor is entitled to an extension of time he shall so inform the Contractor without delay.

Assessment at due date for completion

(4) The Employer's Representative shall not later than 14 days after the due date or extended date for completion of the Works or any Section thereof (and whether or not the Contractor has made any claim for an extension of time) consider all the circumstances known to him at that time and take action similar to that provided for in sub-clause (3) of this Clause.

Should the Employer's Representative consider the Contractor is not entitled to an extension of time he shall so notify the Employer and the Contractor.

Final determination of extension

(5) The Employer's Representative shall within 28 days of the issue of the Certificate of Substantial Completion for the Works or for any Section thereof review all the circumstances of the kind referred to in sub-clause (1) of this Clause and shall finally determine and certify to the Contractor with a copy to the Employer the overall extension of time (if any) to which he considers the Contractor entitled in respect of the Works or the relevant Section.

No such final review of the circumstances shall result in a decrease in any extension of time already granted by the Employer's Representative pursuant to sub-clauses (3) or (4) of this Clause.

Night and Sunday Work

45 Subject to any provisions to the contrary contained in the Contract none of the Permanent and Temporary Works shall be carried out during the night or on Sundays without the permission of the Employer's Representative save when the work is unavoidable or absolutely necessary for the saving of life or property or for the safety of such Works in which case the Contractor shall immediately advise the Employer's Representative.

Provided always that this Clause shall not be applicable in the case of any work which it is customary to carry out outside normal working hours or by rotary or double shifts.

Rate of progress **46** (1) If for any reason which does not entitle the Contractor to an extension of time the rate of progress of the Works or any Section is at any time in the opinion of the Employer's Representative too slow to ensure substantial completion by the time or extended time for completion prescribed by Clauses 43 and 44 as appropriate or the revised time for completion agreed under sub-clause (3) of this Clause the Employer's Representative shall notify the Contractor in writing.

The Contractor shall thereupon take such steps as are necessary and to which the Employer's Representative may consent to expedite the progress so as to substantially complete the Works or such Section by that prescribed time or extended time. The Contractor shall not be entitled to any additional payment for taking such steps.

Permission to work at night or on Sundays (2) If as a result of any action under sub-clause (1) of this Clause the Contractor seeks the Employer's Representative's permission to do any work on Site at night or on Sundays such permission shall not be unreasonably refused.

Provision for accelerated completion (3) If the Contractor is requested by the Employer or the Employer's Representative to complete the Works or any Section within a revised time being less than the time or extended time for completion prescribed by Clauses 43 and 44 as appropriate and the Contractor agrees so to do then any special terms and conditions of payment shall be agreed between the Contractor and the Employer before any such action is taken.

LIQUIDATED DAMAGES FOR DELAY

Liquidated damages for delay in the Substantial Completion of the whole of the Works **47** (1) (a) Where the whole of the Works is not divided into Sections the Appendix to the Form of Tender shall include a sum which represents the Employer's genuine pre-estimate (expressed per week or per day as the case may be) of the damages likely to be suffered by him if the whole of the Works is not substantially completed within the time prescribed by Clause 43 or any extension thereof granted under Clause 44 or by any revision thereof agreed under Clause 46(3) as the case may be.

(b) If the Contractor fails to achieve substantial completion of the whole of the Works within the time so prescribed he shall pay to the Employer the said sum for every week or day (as the case may be) which shall elapse between the date on which the prescribed time expired and the date the whole of the Works is substantially completed.

Provided that if any part of the Works is certified as complete pursuant to Clause 48 before the substantial completion of the whole of the Works the said sum shall be reduced by the proportion which the value of the part so completed bears to the value of the whole of the Works.

Liquidated damages for delay in Substantial Completion where the whole of the Works is divided into Sections (2) (a) Where the Works is divided into Sections (together comprising the whole of the Works) which are required to be substantially completed within particular times as stated in the Appendix to the Form of Tender sub-clause (1) of this Clause shall not apply and the said Appendix shall include a sum in respect of each Section which represents the Employer's genuine preestimate (expressed per week or per day as the case may be) of the damages likely to be suffered by him if that Section is not substantially completed within the time prescribed by Clause 43 or any extension thereof granted under Clause 44 or by any revision thereof agreed under Clause 46(3) as the case may be.

(b) If the Contractor fails to achieve substantial completion of any Section within the time so prescribed he shall pay to the Employer the appropriate stated sum for every week or day (as the case may be) which shall elapse between the date on which the prescribed time expired and the date of substantial completion of that Section.

Provided that if any part of that Section is certified as substantially complete pursuant to Clause 48 before the substantial completion of the whole thereof the appropriate stated sum shall be reduced by the proportion which the value of the part bears to the value of the whole of that Section.

(c) Liquidated damages in respect of two or more Sections may where circumstances so dictate run concurrently.

Damages not a penalty

(3) All sums payable by the Contractor to the Employer pursuant to this Clause shall be paid as liquidated damages for delay and not as a penalty.

Limitation of liquidated damages

(4) (a) The total amount of liquidated damages in respect of the whole of the Works or any Section thereof shall be limited to the appropriate sum stated in the Appendix to the Form of Tender. If no such limit is stated therein then liquidated damages without limit shall apply.

(b) Should there be omitted from the Appendix to the Form of Tender any sum required to be inserted therein either by sub-clause (1)(a) or by sub-clause (2)(a) of this Clause as the case may be or if any such sum is stated to be "nil" then to that extent damages shall not be payable.

Recovery and reimbursement of liquidated damages

(5) The Employer may

(a) deduct and retain the amount of any liquidated damages becoming due under the provisions of this Clause from any sums due or which become due to the Contractor or

(b) require the Contractor to pay such amount to the Employer forthwith.

If upon a subsequent or final review of the circumstances causing delay the Employer's Representative grants a relevant extension or further extension of time the Employer shall no longer be entitled to liquidated damages in respect of the period of such extension.

Any sum in respect of such period which may already have been recovered under this Clause shall be reimbursed forthwith to the Contractor together with interest as provided in Clause 60(7) from the date on which such sums were recovered from the Contractor.

Intervention of alterations etc.

(6) If after liquidated damages have become payable in respect of any part of the Works the Employer's Representative issues an alteration order under Clause 51 or adverse physical conditions or artificial obstructions within the meaning of Clause 12 are encountered or any other situation outside the Contractor's control arises any of which in the opinion of the Employer's Representative results in further delay to that part of the Works

(a) the Employer's Representative shall so notify the Contractor and the Employer in writing and

(b) the Employer's entitlement to liquidated damages in respect of that part of the Works shall be suspended until the Employer's Representative notifies the Contractor and the Employer in writing that the further delay has come to an end.

Such suspension shall not invalidate any entitlement to liquidated damages which accrued before the period of further delay started to run and subject to any subsequent or final review of the circumstances causing delay any monies already deducted or paid as liquidated damages under the provisions of this Clause may be retained by the Employer .

CERTIFICATE OF SUBSTANTIAL COMPLETION

Notification of Substantial Completion

48 (1) When the Contractor considers that

(a) the whole of the Works or

(b) any Section in respect of which a separate time for completion is provided in the Appendix to the Form of Tender

has been substantially completed and has satisfactorily passed any final test that may be prescribed by the Contract or by Statute he may give notice in writing to that effect to the Employer's Representative. Such notice shall be accompanied by an undertaking to finish any outstanding work in accordance with the provisions of Clause 49(1).

Certificate of Substantial Completion

(2) The Employer's Representative shall within 21 days of the date of delivery of such notice either

(a) issue to the Contractor (with a copy to the Employer) a Certificate of Substantial Completion stating the date on which in his opinion the Works were or the Section was substantially completed in accordance with the Contract or

(b) give instructions in writing to the Contractor specifying all the work which in the Employer's Representative's opinion requires to be done by the Contractor before the issue of such certificate.

If the Employer's Representative gives such instructions the Contractor shall be entitled to receive a Certificate of Substantial Completion within 21 days of completion to the satisfaction of the Employer's Representative of the work specified in the said instructions.

Operation and maintenance instructions

(3) Prior to the issue of any Certificate of Substantial Completion under sub-clause (2) of this Clause the Contractor shall provide the Employer with operation and maintenance instructions to enable the Works being taken over by the Employer to be operated maintained dismantled re-assembled and adjusted satisfactorily.

Premature use by Employer

(4) If any substantial part of the Works has been occupied or used by the Employer other than as provided in the Contract the Contractor may request in writing and the Employer's Representative shall issue a Certificate of Substantial Completion in respect thereof. Such certificate shall take effect from the date of delivery of the Contractor's request and upon the issue of such certificate the Contractor shall be deemed to have undertaken to complete any outstanding work in that part of the Works during the Defects Correction Period.

Substantial Completion of other parts of the Works

(5) If the Employer's Representative considers that any part of the Works has been substantially completed and has passed any final test that may be prescribed by the Contract he may issue a Certificate of Substantial Completion in respect of that part of the Works before completion of the whole of the Works and upon the issue of such certificate the Contractor shall be deemed to have undertaken to complete any outstanding work in that part of the Works during the Defects Correction Period.

Reinstatement of ground

(6) A Certificate of Substantial Completion given in respect of any Section or part of the Works before completion of the whole shall not be deemed to certify completion of any ground or surfaces requiring reinstatement unless such certificate shall expressly so state.

OUTSTANDING WORK AND DEFECTS

Work outstanding

49 (1) The undertaking to be given under Clause 48(1) may by agreement between the Employer's Representative and the Contractor specify a time or times within which the outstanding work shall be completed. If no such times are specified any outstanding work shall be completed as soon as practicable during the Defects Correction Period.

Carrying out of work of repair etc.

(2) The Contractor shall deliver up to the Employer the Works and each Section and part thereof at or as soon as practicable after the end of the relevant Defects Correction Period in the condition required by the Contract (fair wear and tear excepted) to the satisfaction of the Employer's Representative.

To this end the Contractor shall repair amend reconstruct rectify and make good defects of whatever nature notified to him in writing by the Employer's Representative during the relevant Defects Correction Period or within 14 days after its expiry as a result of an inspection made by or on behalf of the Employer's Representative prior to its expiry.

Cost of work of repair etc.

(3) All work required under sub-clause (2) of this Clause shall be carried out by the Contractor at his own expense if in the Employer's Representative's opinion it is necessary due to failure by the Contractor to comply with any of his obligations under the Contract. In any other event the value of such work shall be ascertained and paid for as if it were additional work.

Remedy on Contractor's failure to carry out work required

(4) If the Contractor fails to do any such work as aforesaid the Employer shall be entitled to carry out that work by his own workpeople or by other contractors and if it is work which the Contractor should have carried out at his own expense the Employer shall be entitled to recover the cost thereof from the Contractor and may deduct the same from any monies that are or may become due to the Contractor.

Contractor to search

50 The Contractor shall if required by the Employer's Representative in writing carry out such searches tests or trials as may be necessary to determine the cause of any defect imperfection or fault.

Unless the defect imperfection or fault is one for which the Contractor is liable under the Contract the cost of the work carried out by the Contractor as aforesaid shall be borne by the Employer. If the defect imperfection or fault is one for which the Contractor is liable the cost of the work carried out as aforesaid shall be borne by the Contractor and he shall in such case repair rectify and make good the defect imperfection or fault at his own expense in accordance with Clause 49.

ALTERATIONS AND ADDITIONAL PAYMENTS

Ordered alterations to Employer's Requirements

51 (1) The Employer's Representative shall have power after consultation with the Contractor's Representative to alter the Employer's Requirements. Such alterations may include additions and/or omissions and may be ordered at any time up to the end of the Defects Correction Period for the whole of the Works.

Ordered alterations to be in writing

(2) All alterations shall be ordered in writing but the provisions of Clause 2(5) in respect of oral instructions shall apply.

Alterations proposed by the Contractor

(3) The Contractor may at any time submit to the Employer's Representative proposals to alter the Employer's Requirements. If such proposals are accepted (whether as submitted or with modifications) the Employer's Representative shall issue an appropriate order under sub-clause (1) of this Clause.

Alteration not to affect Contract

(4) No alteration ordered under this Clause shall in any way vitiate or invalidate the Contract but the fair and reasonable value (if any) of all such alterations shall be taken into account in ascertaining the amount of the Contract Price except to the extent that such alteration is necessitated by the Contractor's default.

Valuation of ordered alterations

52 (1) If requested by the Employer's Representative the Contractor shall submit his quotation for the work as altered and his estimate of any delay. Wherever possible the value and delay consequences (if any) of each alteration shall be agreed before the order is issued or before work starts.

Sharing the consequences

(2) Where the altered work results from a Contractor's proposal pursuant to Clause 51(3) the consequences (if any) as to delay acceleration or valuation shall be shared between the Employer and the Contractor having regard to the particular circumstances of each case.

(3) In all other cases the valuation of alterations ordered by the Employer's Representative in accordance with Clause 51 shall be ascertained as follows

(a) As soon as possible after receipt of the alteration order the Contractor shall submit to the Employer's Representative

(i) his quotation for any extra or substituted works necessitated by the alteration having due regard to any rates or prices included in the Contract and

(ii) his estimate of any delay occasioned thereby and

(iii) his estimate of the cost of any such delay.

(b) Within 14 days of receiving the said submissions the Employer's Representative shall

(i) accept the same or

(ii) negotiate with the Contractor thereon.

(c) Upon reaching agreement with the Contractor the Contract Price shall be amended accordingly.

(d) Failing agreement the Employer's Representative shall notify the Contractor of what in his opinion is a fair and reasonable valuation and thereafter shall make such interim valuations for the purposes of Clause 60 as may be appropriate.

Daywork

(4) The Employer's Representative may if in his opinion it is necessary or desirable order in writing that any additional or substituted work shall be executed on a daywork basis in accordance with the provisions of Clause 56(3).

Additional payments **53**

(1) If the Contractor intends to claim any additional payment pursuant to any Clause of these Conditions other than Clause 52(1) he shall give notice in writing of his intention to the Employer's Representative as soon as may be reasonable and in any event within 28 days after the happening of the events giving rise to the claim.

Upon the happening of such events the Contractor shall keep such contemporary records as may reasonably be necessary to support any claim he may subsequently wish to make.

(2) Without necessarily admitting the Employer's liability the Employer's Representative may upon receipt of a notice under this Clause instruct the Contractor to keep such contemporary records or further contemporary records as the case may be as are reasonable and may be material to the claim of which notice has been given and the Contractor shall keep such records.

The Contractor shall permit the Employer's Representative to inspect all records kept pursuant to this Clause and shall supply him with copies thereof as and when the Employer's Representative shall so instruct.

(3) After the giving of a notice to the Employer's Representative under this Clause the Contractor shall as soon as is reasonable in all the circumstances send to the Employer's Representative a first interim account giving full and detailed particulars of the amount claimed to that date and of the grounds upon which the claim is based.

Thereafter at such intervals as the Employer's Representative may reasonably require the Contractor shall send to the Employer's Representative further up to date accounts giving the accumulated total of the claim and any further grounds upon which it is based.

(4) If the Contractor fails to comply with any of the provisions of this Clause in respect of any claim which he shall seek to make then the Contractor shall be entitled to payment in respect thereof only to the extent that the Employer's Representative has not been prevented from or substantially prejudiced by such failure in investigating the said claim.

(5) The Contractor shall be entitled to have included in any interim payment certified by the Employer's Representative pursuant to Clause 60 such amount in respect of any claim as the Employer's Representative may consider due to the Contractor provided that the Contractor shall have supplied sufficient particulars to enable the Employer's Representative to determine the amount due.

If such particulars are insufficient to substantiate the whole of the claim the Contractor shall be entitled to payment in respect of such part of the claim as the particulars may substantiate to the satisfaction of the Employer's Representative.

MATERIALS AND CONTRACTOR'S EQUIPMENT

Non-removal of materials and Contractor's Equipment

54 (1) No Contractor's Equipment Temporary Works materials for Temporary Works or other goods or materials owned by the Contractor and brought on to the Site for the purposes of the Contract shall be removed without the written consent of the Employer's Representative which consent shall not unreasonably be withheld.

Liability for loss or damage to Contractor's Equipment

(2) The Employer shall not at any time be liable save as mentioned in Clauses 20(2) and 63(5) for the loss of or damage to any Contractor's Equipment Temporary Works goods or materials.

Disposal of Contractor's Equipment

(3) If the Contractor fails to remove any of the said Contractor's Equipment Temporary Works goods or materials as required by Clause 33 within such reasonable time after completion of the Works as the Employer's Representative may allow then the Employer may sell or otherwise dispose of such items. From the proceeds of the sale of any such items the Employer shall be entitled to retain any costs or expenses incurred in connection with their sale and disposal before paying the balance (if any) to the Contractor.

Vesting of goods and materials not on Site

(4) With a view to securing payment under Clause 60(1)(c) the Contractor may (and shall if the Employer's Representative so directs) transfer to the Employer the property in goods and materials listed in the Appendix to the Form of Tender or as subsequently agreed between the Contractor and the Employer before the same are delivered to the Site provided that the goods and materials

(a) have been manufactured or prepared and are substantially ready for incorporation in the Works and

(b) are the property of the Contractor or the contract for the supply of the same expressly provides that the property therein shall pass unconditionally to the Contractor upon the Contractor taking the action referred to in sub-clause (5) of this Clause and

(c) have been marked and set aside in accordance with sub-clause (5) of this Clause.

Action by Contractor

(5) The intention of the Contractor to transfer the property in any goods or materials to the Employer in accordance with this Clause shall be evidenced by the Contractor taking or causing the supplier of those goods or materials to take the following actions.

(a) Provide to the Employer's Representative documentary evidence that the property of the said goods or materials has vested in the Contractor.

(b) Suitably mark or otherwise plainly identify the goods and materials so as to show that their destination is the Site that they are the property of the Employer and (where they are not stored at the premises of the Contractor) to whose order they are held.

(c) Set aside and store the said goods and materials so marked and identified to the satisfaction of the Employer's Representative.

(d) Send to the Employer's Representative a schedule listing and giving the value of every item of the goods and materials so set aside and stored and inviting him to inspect them.

Vesting in the Employer

(6) Upon the Employer's Representative approving in writing the transfer in ownership of any goods and materials for the purposes of this Clause they shall vest in and become the absolute property of the Employer and thereafter shall be in the possession of the Contractor for the sole purpose of delivering them to the Employer and incorporating them in the Works and shall not be within the ownership control or disposition of the Contractor. Provided always that

(a) approval by the Employer's Representative for the purposes of this Clause or any payment certified by him in respect of goods and materials pursuant to Clause 60 shall be without prejudice to the exercise of any power of the Employer's Representative contained in this Contract to reject any goods or materials which are not in accordance with the provisions of the Contract and upon any such rejection the property in the rejected goods or materials shall immediately re-vest in the Contractor and

(b) the Contractor shall be responsible for any loss or damage to such goods or materials and for the cost of storing handling and transporting the same and shall effect such additional insurance as may be necessary to cover the risk of such loss or damage from any cause.

Lien on goods and materials

(7) Neither the Contractor nor a sub-contractor nor any other person shall have a lien on any goods or materials which have vested in the Employer under sub-clause (6) of this Clause for any sum due to the Contractor sub-contractor or other person and the Contractor shall take all steps reasonably necessary to ensure that the title of the Employer and the exclusion of any such lien are brought to the notice of sub-contractors and other persons dealing with such goods or materials.

Delivery to the Employer of vested goods or materials

(8) Upon cessation of the employment of the Contractor under this Contract before the completion of the Works whether as a result of the operation of Clause 63 64 or 65 or otherwise the Contractor shall deliver to the Employer any goods or materials the property in which has vested in the Employer by virtue of sub-clause (6) of this Clause and if he fails to do so the Employer may enter any premises of the Contractor or of any sub-contractor and remove such goods and materials and recover the cost of doing so from the Contractor.

Incorporation in sub-contracts

(9) The Contractor shall incorporate provisions equivalent to those provided in this Clause in every sub-contract in which provision is to be made for the payment in respect of goods or materials before the same have been delivered to the Site.

MEASUREMENT

Quantities **55**

(1) Where the Contract includes any bill of quantities schedule of works or Provisional Sums the quantities set out therein shall be deemed to be the estimated quantities of the work but they are not to be taken as the actual or correct quantities of work to be constructed by the Contractor in fulfilment of his obligations under the Contract.

(2) No error in description in or omission from any such bill of quantities schedule of works or Provisional Sum or in any schedule of rates shall vitiate the Contract nor release the Contractor from any of his obligations or liabilities thereunder.

Attending for measurement **56**

(1) If and to the extent that it becomes necessary to measure any part of the Works either the Contractor's or the Employer's Representative shall give reasonable notice in writing to the other requiring him to attend or send a qualified agent to assist in making such measurement.

(2) Should either the Contractor's or the Employer's Representative not attend or neglect or omit to send such agent when so requested then the measurement made by the other or approved by him shall be taken to be the correct measurement of the work.

Daywork

(3) Where any work is carried out on a daywork basis the Contractor shall be paid for such work under the conditions and at the rates and prices set out in the daywork schedule included in the Contract or failing the inclusion of a daywork schedule he shall be paid at the rates and prices and under the conditions contained in the "Schedule of Dayworks carried out incidental to Contract Work" issued by the Civil Engineering Contractors Association current at the date of the execution of the daywork.

The Contractor shall furnish to the Employer's Representative such records receipts and other documentation as may be necessary to prove amounts paid and/or costs incurred. Such returns shall be in the form and delivered at the times the Employer's Representative shall direct and shall be agreed within a reasonable time.

Before ordering materials the Contractor shall if so required submit to the Employer's Representative quotations for the same for his approval.

57 Not Used.

PROVISIONAL SUM AND PRIME COST ITEMS

Use of Provisional Sum and Prime Cost Items

58 (1) The Contractor shall not commence work on any Provisional Sum or Prime Cost Item until he has secured the consent thereto of the Employer's Representative which consent shall not unreasonably be withheld.

Valuation and payment

(2) Provisional Sums and Prime Cost Items shall be valued and paid for in accordance with Clause 52 or as the Contract otherwise provides. The percentage to be used for overheads and profit in adjusting the Prime Cost element of any such item shall be the figure stated therefore in the Appendix to the Form of Tender.

59 Not used.

CERTIFICATES AND PAYMENT

Interim statements

60 (1) The Contractor shall submit to the Employer's Representative at such times and in such form as the Contract may prescribe statements showing the amounts which in the Contractor's opinion are due under the Contract and where appropriate showing separately

(a) each amount expended against Provisional Sum and Prime Cost Items

(b) a list of any goods or materials delivered to the Site for but not yet incorporated in the Permanent Works and their value

(c) a list of any goods or materials which have not yet been delivered to the Site but of which the property has vested in the Employer pursuant to Clause 54 and their value and

(d) the estimated amounts to which the Contractor considers himself entitled in connection with all other matters for which provision is made under the Contract

unless in the opinion of the Contractor such values and amounts together will not justify the issue of an interim certificate.

Interim payment

(2) Within 25 days of the date of delivery of the Contractor's statement to the Employer's Representative in accordance with sub-clause (1) of this Clause the Employer's Representative shall either

(a) where a payment schedule is included in the Contract certify

(i) when the progress of the Works or any part thereof has reached the state required for the payment of the relevant interim or stage payment and

(ii) the amount of such payment less a retention as provided in sub-clause (5) of this Clause or

(b) where no such schedule is included in the Contract certify

(i) the amount which in the opinion of the Employer's Representative on the basis of the Contractor's statement is due to the Contractor other than on account of sub-clauses (1)(b) and (1)(c) of this Clause less a retention as provided in sub-clause (5) of this Clause and

(ii) such amounts (if any) as the Employer's Representative may consider proper (but in no case exceeding the percentage of the value stated in the Appendix to the Form of Tender) in respect of sub-clauses (1)(b) and (1)(c) of this Clause

and the Employer shall pay to the Contractor such amounts as become due thereby after deducting any previous payments on account.

The payments become due on certification with the final date for payment being 28 days after the date of delivery of the Contractor's statement.

Minimum amount of certificate

(3) Until the whole of the Works has been certified as substantially complete in accordance with Clause 48 the Employer's Representative shall not be bound to issue an interim certificate for a sum less than that stated in the Appendix to the Form of Tender but thereafter he shall be bound to do so and the certification and payment of amounts due to the Contractor shall be in accordance with the time limits contained in this Clause.

Final account

(4) Not later than 3 months after the date of the Defects Correction Certificate the Contractor shall submit to the Employer's Representative a statement of the final account and supporting documentation showing in detail the value in accordance with the Contract of the Works carried out together with all further sums which the Contractor considers to be due to him under the Contract up to the date of the Defects Correction Certificate.

Within 3 months after receipt of this final account and of all information reasonably required for its verification the Employer's Representative shall issue a certificate stating the amount which in his opinion is finally due under the Contract from the Employer to the Contractor or from the Contractor to the Employer (as the case may be) up to the date of the Defects Correction Certificate and after giving credit to the Employer for all amounts previously paid by the Employer and for all sums to which the Employer is entitled under the Contract.

Such amount shall subject to Clause 47 be paid to or by the Contractor as the case may require. The payment becomes due on certification. The final date for payment is 28 days later.

Retention

(5) The retention to be made pursuant to sub-clause (2) of this Clause shall be the difference between

(a) an amount calculated at the rate indicated in and up to the limit set out in the Appendix to the Form of Tender upon the amount due to the Contractor other than on account of sub-clauses (1)(b) and (1)(c) of this Clause and

(b) any payment which shall have become due under sub-clause (6) of this Clause.

Payment of retention

(6) (a) Upon the issue of a Certificate of Substantial Completion in respect of any Section or part of the Works there shall become due to the Contractor one half of such proportion of the retention money deductible to date under sub-clause (5)(a) of this Clause as the value of the Section or part bears to the value of the whole of the Works completed to date as certified under sub-clause (2) of this Clause. The final date for payment shall be 14 days after the date of issue of the relevant Certificate of Substantial Completion.

The total of the payments which shall become due under this sub-clause shall in no event exceed one half of the limit of retention set out in the Appendix to the Form of Tender.

(b) Upon issue of the Certificate of Substantial Completion in respect of the whole of the Works there shall become due to the Contractor one half of the retention money calculated in accordance with sub-clause (5)(a)(or the balance thereof over and above such payments already made pursuant to sub-clause (6)(a) of this Clause). The final date for payment shall be 14 days after the date of issue of the Certificate of Substantial Completion for the Whole of the Works

(c) At the end of the Defects Correction Period or if more than one the last of such periods there shall become due to the Contractor the remainder of the retention money. Within 10 days of the date of the end of the said period the Employers Representative shall certify the amount due. The final date for payment shall be 14 days after the end of the relevant Defects Correction Period.

Provided that if at that time there remains to be carried out by the Contractor any outstanding work referred to under Clause 48 or any work ordered pursuant to Clauses 49 or 50 the Employer may by giving due notice under Clause 60(10) withhold payment until the completion of such work of so much of the said remainder as shall in the opinion of the Employer's Representative represent the cost of the work remaining to be carried out.

Interest on overdue payments

(7) In the event of

(a) failure by the Employer's Representative to certify or the Employer to make payment in accordance with sub-clauses (2) (4) or (6) of this Clause or

(b) any decision of an adjudicator or any finding of an arbitrator to such effect

the Employer shall pay to the Contractor interest compounded monthly for each day on which any payment is overdue or which should have been certified and paid at a rate equivalent to 2% per annum above the base lending rate of the bank specified in the Appendix to the Form of Tender.

If in an arbitration pursuant to Clause 66 the arbitrator holds that any sum or additional sum should have been certified by a particular date in accordance with the aforementioned sub-clauses but was not so certified this shall be regarded for the purposes of this sub-clause as a failure to certify such sum or additional sum. Such sum or additional sum shall be regarded as overdue for payment 28 days after the date by which the arbitrator holds that the Employer's Representative should have certified the sum or if no such date is identified by the arbitrator shall be regarded as overdue for payment from the date of the Certificate of Substantial Completion for the whole of the Works.

Correction and withholding of certificates

(8) The Employer's Representative shall have power to omit from any certificate the value of any work done goods or materials supplied or services rendered with which he may for the time being be dissatisfied and for that purpose or for any other reason which to him may seem proper may by any certificate delete correct or modify any sum or statement of fact previously certified by him.

Certificates and payment notices

(9) Every certificate issued by the Employer's Representative pursuant to this Clause shall be sent to the Employer and on the Employer's behalf to the Contractor. By this certificate the Employer shall give notice to the Contractor specifying the amount (if any) of the payment proposed to be made and the basis on which it was calculated.

Notice of intention to withhold payment

(10) Where a payment under Clause 60(2), (4) or (6) is to differ from that certified or the Employer is to withhold payment after the final date for payment of a sum due under the Contract the Employer shall notify the Contractor in writing not less than one day before the final date for payment specifying the amount proposed to be withheld and the ground for withholding payment or if there is more than one ground each ground and the amount attributable to it.

Defects Correction Certificate **61**

(1) At the end of the Defects Correction Period or where there is more than one such period at the end of the last of such periods and when all outstanding work referred to under Clause 48 and all work of repair amendment reconstruction rectification and making good of defects imperfections shrinkages and other faults referred to under Clauses 49 and 50 has been completed the Employer's Representative shall issue to the Employer (with a copy to the Contractor) a Defects Correction Certificate stating the date on which the Contractor has completed his obligations to construct and complete the Works to the Employer's Representative's satisfaction.

Unfulfilled obligations

(2) The issue of the Defects Correction Certificate shall not be taken as relieving either the Contractor or the Employer from any liability the one towards the other arising out of or in any way connected with the performance of their respective obligations under the Contract.

Manuals and drawings

(3) Prior to the issue of the Defects Correction Certificate the Contractor shall

(a) submit to the Employer's Representative for his approval one set of draft operation and maintenance manuals together with as-constructed record drawings in sufficient detail to enable the Employer to operate maintain dismantle reassemble and adjust the Permanent Works and

(b) once the Employer's Representative has given approval thereto supply to the Employer three sets of the finally approved manuals and drawings.

REMEDIES AND POWERS

Urgent repairs **62** If in the opinion of the Employer's Representative any remedial or other work or repair is urgently necessary by reason of any accident or failure or other event occurring to in or in connection with the Works or any part thereof either during the carrying out of the Works or during the Defects Correction Period the Employer's Representative shall so inform the Contractor with confirmation in writing.

Thereafter if the Contractor is unable or unwilling to carry out such work or repair at once the Employer may himself carry out the said work or repair using his own or other workpeople.

If the work or repair so carried out by the Employer is work which in the opinion of the Employer's Representative the Contractor was liable to carry out at his own expense under the Contract all costs and charges properly incurred by the Employer in so doing shall on demand be paid by the Contractor to the Employer or may be deducted by the Employer from any monies due or which may become due to the Contractor.

Frustration **63** (1) If any circumstance outside the control of both parties arises during the currency of the Contract which renders it impossible or illegal for either party to fulfil his contractual obligations the Works shall be deemed to be abandoned upon the service by one party upon the other of written notice to that effect.

War clause

(2) If during the currency of the Contract there is an outbreak of war (whether war is declared or not) in which Great Britain is engaged on a scale involving general mobilisation of the armed forces of the Crown

(a) the Contractor shall for a period of 28 days reckoned from midnight on the date that the order for general mobilisation is given continue so far as is physically possible to execute the Works in accordance with the Contract and

(b) if substantial completion of the whole of the Works is not achieved before the said period of 28 days has expired the Works shall thereupon be deemed to be abandoned unless the parties otherwise agree.

Removal of Contractor's Equipment

(3) Upon abandonment of the Works pursuant to sub-clauses (1) or (2)(b) of this Clause the Contractor shall with all reasonable dispatch remove from the Site all Contractor's Equipment.

In the event of any failure so to do the Employer shall have like powers to those contained in Clause 54(3) to dispose of any Contractor's Equipment.

Payment on abandonment

(4) Upon abandonment of the Works pursuant to sub-clauses (1) or (2)(b) of this Clause the Employer shall pay the Contractor (in so far as such amounts or items have not already been covered by payments on account made to the Contractor) the Contract value of all work carried out prior to the date of abandonment and in addition

(a) the amounts payable in respect of any preliminary items so far as the work or service comprised therein has been carried out or performed and a proper proportion of any such items which have been partially carried out or performed

(b) the cost of materials or goods reasonably ordered for the Works which have been delivered to the Contractor or of which the Contractor is legally liable to accept delivery (such materials or goods becoming the property of the Employer upon such payment being made to the Contractor)

(c) a sum being the amount of any expenditure reasonably incurred by the Contractor in the expectation of completing the whole of the Works insofar as such expenditure has not been recovered by any other payments referred to in this sub-clause and

(d) the reasonable cost of removal under sub-clause (3) of this Clause.

To this end and without prejudice to the provisions of sub-clause (5) of this Clause the provisions of Clause 60(4) shall apply to this sub-clause as if the date of abandonment was the date of issue of the Defects Correction Certificate.

Works substantially completed

(5) If upon abandonment of the Works any Section or part of the Works has been substantially completed in accordance with Clause 48 or is completed so far as to be useable then in connection therewith

(a) the Contractor may at his discretion and in lieu of his obligations under Clauses 49 and 50 allow against the sum due to him pursuant to sub-clause (4) of this Clause the cost (calculated as at the date of abandonment) of repair rectification and making good for which he would have been liable under the said Clauses had they continued to be applicable and

(b) the Employer shall not be entitled to withhold payment under Clause 60(6)(c) of the second half of the retention money or any part thereof except such sum as the Contractor may allow under the provisions of the last preceding paragraph.

Contract to continue in force

(6) Save as aforesaid the Contract shall continue to have full force and effect.

Failure to pay the Contractor

64 (1) In the event of failure by the Employer's Representative to certify or the Employer to pay the Contractor the amount due (subject to any deduction that the Employer is entitled to make under the Contract) in accordance with Clause 60 within 21 days after the expiry of the time therein stated then the Contractor may after giving 7 days notice in writing to the Employer (with a copy to the Employer's Representative) suspend work or reduce the rate of work.

Resumption of work

(2) Upon payment by the Employer of the amount due (including interest pursuant to Clause 60(7)) the Contractor shall resume normal working as soon as is reasonably possible.

Delay and extra cost

(3) If the Contractor suffers delay and/or incurs additional cost as a result of suspending or reducing the rate of work under sub-clause (1) of this Clause the Employer's Representative shall determine

(a) any extension of time to which the Contractor is entitled under Clause 44 and

(b) subject to Clause 53 the amount of such additional cost.

Default of Employer

(4) In the event of

(a) failure by the Employer's Representative to certify or the Employer to pay the Contractor the amount due (subject to any deduction that the Employer is entitled to make under the Contract) in accordance with Clause 60 within 56 days after the expiry of the time therein stated

or if the Employer

(b) assigns or attempts to assign the Contract or any part thereof or any benefit or interest thereunder without the prior written consent of the Contractor or

(c) (i) becomes bankrupt or presents his petition in bankruptcy or

(ii) has a receiving order or administration order made against him or

(iii) makes an arrangement with or an assignment in favour of his creditors or

(iv) agrees to perform the Contract under a committee of inspection of his creditors or

(v) (being a corporation) has a receiver or administrator appointed or goes into liquidation (other than a voluntary liquidation for the purposes of amalgamation or reconstruction) or

(d) has an execution levied on his goods which is not stayed or discharged within 28 days

then the Contractor may after giving 7 days notice in writing to the Employer specifying the default terminate his employment under the Contract without thereby avoiding the Contract or releasing the Employer from any of his obligations or liabilities under the Contract.

Provided that the Contractor may extend the period of notice to give the Employer an opportunity to remedy his default.

Removal of Contractor's Equipment

(5) Upon expiry of the 7 days notice referred to in sub-clause (4) of this Clause and notwithstanding the provisions of Clause 54 the Contractor shall with all reasonable despatch remove from the Site all Contractor's Equipment.

Payment upon termination

(6) Upon termination of the Contractor's employment pursuant to sub-clause (4) of this Clause the Employer shall be under the same obligations with regard to payment as if the Contract had been abandoned under the provisions of Clause 63.

Provided that in addition to payments specified under Clause 63(4) the Employer shall pay to the Contractor

(a) any amount to which the Contractor is entitled under sub-clause (3) of this Clause and

(b) the amount of any loss or damage to the Contractor arising from or as a consequence of such termination.

Default of Contractor **65** (1) In the event that the Contractor

(a) assigns or attempts to assign the Contract or any part thereof or any benefit or interest thereunder without the prior written consent of the Employer or

(b) is in breach of Clause 4(1) or

(c) (i) becomes bankrupt or presents his petition in bankruptcy or

(ii) has a receiving order or administration order made against him or

(iii) makes an arrangement with or an assignment in favour of his creditors or

(iv) agrees to carry out the Contract under a committee of inspection of his creditors or

(v) (being a corporation) has a receiver or administrator appointed or goes into liquidation (other than a voluntary liquidation for the purposes of amalgamation or reconstruction) or

(d) has an execution levied on his goods which is not stayed or discharged within 28 days

or if the Employer's Representative certifies in writing to the Employer with a copy to the Contractor that in his opinion the Contractor

(e) has abandoned the Contract without due cause or

(f) without reasonable excuse has failed to commence the Works in accordance with Clause 41 or

(g) has suspended the progress of the Works without due cause for 14 days after receiving from the Employer's Representative written notice to proceed or

(h) has failed to remove goods or materials from the Site or to pull down and replace work for 14 days after receiving from the Employer's Representative written notice that the said goods materials or work has been condemned and rejected by the Employer's Representative or

(j) despite previous warnings by the Employer's Representative in writing is failing to proceed with the Works with due diligence or is otherwise persistently or fundamentally in breach of his obligations under the Contract then the Employer may after giving 7 days notice in writing to the Contractor specifying the default enter upon the Works and any other parts of the Site provided by the Employer and expel the Contractor therefrom without thereby avoiding the Contract or releasing the Contractor from any of his obligations or liabilities under the Contract.

Where a notice of termination is given pursuant to a certificate issued by the Employer's Representative under this sub-clause it shall be given as soon as is reasonably possible after receipt of the certificate.

Provided that the Employer may extend the period of notice to give the Contractor an opportunity to remedy the default.

Completing the Works

(2) Where the Employer has entered upon the Works and any other parts of the Site as set out in sub-clause (1) of this Clause he may

(a) complete the Works himself or

(b) employ any other contractor to complete the Works

and in either case may use for such completion any of the Contractor's equipment Temporary Works goods and materials on any part of the Site. For this purpose the Employer shall continue to enjoy his right under Clause 7(1)(b) and to copy and use all drawings specifications and other documents made by or on behalf of the Contractor.

The Employer may at any time sell any of the said Contractor's Equipment Temporary Works and unused goods and materials and apply the proceeds of sale in or towards the satisfaction of any sums due or which may become due to him from the Contractor under the Contract.

Assignment to Employer

(3) Where the Employer has entered upon the Works and any other parts of the Site as hereinbefore provided the Contractor shall if so instructed by the Employer's Representative in writing within 7 days of such entry assign to the Employer the benefit of any agreement which the Contractor may have entered into for the supply of any goods or materials and/or for the carrying out of any work for the purposes of the Contract.

Valuation at date of termination

(4) As soon as may be practicable after any such entry and expulsion by the Employer the Employer's Representative shall fix and determine as at the time of such entry and expulsion

(a) the amount (if any) which has been reasonably earned by or would reasonably accrue to the Contractor in respect of work actually done by him under the Contract and

(b) the value of any unused or partially used goods and materials and any Contractor's Equipment and Temporary Works which had been deemed to become the property of the Employer under Clause 54

and shall certify accordingly.

The said determination may be carried out *ex parte* or by or after reference to the parties or after such investigation or enquiry as the Employer's Representative may think fit to make or institute.

Payment after termination

(5) (a) If the Employer enters and expels the Contractor under this Clause he shall not be liable to pay the Contractor any money under the Contract (whether in respect of amounts certified by the Employer's Representative or otherwise) unless or until the Employer's Representative certifies that an amount is due to the Contractor under sub-clause (b).

(b) The Employer's Representative shall certify the difference between

(i) such sum as would have been due to the Contractor if he had completed the Works together with any proceeds of sale under sub-clause 2 of this Clause and

(ii) the costs of completing the Works (whether or not the Works are completed under a separate contract) damages for delay (if any) and all other expenses properly incurred by the Employer.

(c) Such difference as is certified by the Employer's Representative in sub-clause (b) shall be a debt due to the Employer or Contractor as the case may be.

(d) If the Employer's Representative is satisfied at any time prior to the completion of the Works that such sum as calculated under sub-clause (b)(ii) exceeds such sum as calculated under sub-clause (b)(i) he may issue an interim certificate to that effect notwithstanding that the Works have not been completed and such interim certificate shall be considered a debt due from the Contractor to the Employer.

(e) Every certificate issued by the Employer's Representative pursuant to this Clause shall be sent to the Employer and at the same time copied to the Contractor with such detailed explanation as may be necessary.

SETTLEMENT OF DISPUTES

Avoidance of disputes

66 (1) In order to overcome where possible the causes of disputes and in those cases where disputes are likely still to arise to facilitate their clear definition and early resolution (whether by agreement or otherwise) the following procedure shall apply for the avoidance and settlement of disputes.

Matters of dissatisfaction

(2) If at any time

(a) the Contractor is dissatisfied with any act or instruction of the Employer's Representative or any other person responsible to him or

(b) the Employer or the Contractor is dissatisfied with any decision opinion instruction direction certificate or valuation of the Employer's Representative or with any other matter arising under or in connection with the Contract or the carrying out of the Works

the matter of dissatisfaction shall be referred to the Employer's Representative who shall notify his written decision to the Employer and the Contractor within one month of the reference to him.

Disputes

(3) The Employer and the Contractor agree that no matter shall constitute nor be said to give rise to a dispute unless and until in respect of that matter

(a) the time for the giving of a decision by the Employer's Representative on a matter of dissatisfaction under Clause 66(2) has expired or the decision given is unacceptable or has not been implemented and in consequence the Employer or the Contractor has served on the other and on the Employer's Representative a notice in writing (hereinafter called the Notice of Dispute) or

(b) an adjudicator has given a decision on a dispute under Clause 66(6) and the Employer or the Contractor is not giving effect to the decision, and in consequence the other has served on him and the Employer's Representative a Notice of Dispute

and the dispute shall be that stated in the Notice of Dispute. For the purposes of all matters arising under or in connection with the Contract or the carrying out of the Works the word "dispute" shall be construed accordingly and shall include any difference.

(4) (a) Notwithstanding the existence of a dispute following the service of a Notice under Clause 66(3) and unless the Contract has already been determined or abandoned the Employer and the Contractor shall continue to perform their obligations.

(b) The Employer and the Contractor shall give effect forthwith to every decision of

(i) the Employer's Representative on a matter of dissatisfaction given under Clause 66(2) and

(ii) the adjudicator on a dispute given under Clause 66(6)

unless and until that decision is revised by agreement of the Employer and Contractor or pursuant to Clause 66.

Conciliation

(5) (a) The Employer or the Contractor may at any time before service of a Notice to Refer to arbitration under Clause 66(9) by notice in writing seek the agreement of the other for the dispute to be considered under the Institution of Civil Engineers' Conciliation Procedure (1999) or any amendment or modification thereof being in force at the date of such notice.

(b) If the other party agrees to this procedure any recommendation of the conciliator shall be deemed to have been accepted as finally determining the dispute by agreement so that the matter is no longer in dispute unless a Notice of Adjudication under Clause 66(6) or a Notice to Refer to arbitration under Clause 66(9) has been served in respect of that dispute not later than 1 month after receipt of the recommendation by the dissenting party.

Adjudication

(6) (a) The Employer and the Contractor each has the right to refer any dispute as to a matter under the Contract for adjudication and either party may give notice in writing (hereinafter called the Notice of Adjudication) to the other at any time of his intention so to do. The adjudication shall be conducted under the Institution of Civil Engineers' Adjudication Procedure (1997) or any amendment or modification thereof being in force at the time of the said Notice.

(b) Unless the adjudicator has already been appointed he is to be appointed by a timetable with the object of securing his appointment and referral of the dispute to him within 7 days of such notice.

(c) The adjudicator shall reach a decision within 28 days of referral or such longer period as is agreed by the parties after the dispute has been referred.

(d) The adjudicator may extend the period of 28 days by up to 14 days with the consent of the party by whom the dispute was referred.

(e) The adjudicator shall act impartially.

(f) The adjudicator may take the initiative in ascertaining the facts and the law.

(7) The decision of the adjudicator shall be binding until the dispute is finally determined by legal proceedings or by arbitration (if the contract provides for arbitration or the parties otherwise agree to arbitration) or by agreement.

(8) The adjudicator is not liable for anything done or omitted in the discharge or purported discharge of his functions as adjudicator unless the act or omission is in bad faith and any employee or agent of the adjudicator is similarly not liable.

Arbitration

(9) (a) All disputes arising under or in connection with the Contract or the carrying out of the Works other than failure to give effect to a decision of an adjudicator shall be finally determined by reference to arbitration. The party seeking arbitration shall serve on the other party a notice in writing (called the Notice to Refer) to refer the dispute to arbitration.

(b) Where an adjudicator has given a decision under Clause 66(6) in respect of the particular dispute the Notice to Refer must be served within three months of the giving of the decision otherwise it shall be final as well as binding.

Appointment of arbitrator

(10) (a) The arbitrator shall be a person appointed by agreement of the parties.

President or Vice-President to act

(b) If the parties fail to appoint an arbitrator within one month of either party serving on the other party a notice in writing (hereinafter called the Notice to Concur) to concur in the appointment of an arbitrator the dispute shall be referred to a person to be appointed on the application of either party by the President for the time being of the Institution of Civil Engineers.

(c) If an arbitrator declines the appointment or after appointment is removed by order of a competent court or is incapable of acting or dies and the parties do not within one month of the vacancy arising fill the vacancy then either party may apply to the President for the time being of the Institution of Civil Engineers to appoint another arbitrator to fill the vacancy.

(d) In any case where the President for the time being of the Institution of Civil Engineers is not able to exercise the functions conferred on him by this Clause the said functions shall be exercised on his behalf by a Vice-President for the time being of the said Institution.

Arbitration – procedure and powers

(11) (a) Any reference to arbitration under this Clause shall be deemed to be a submission to arbitration within the meaning of the Arbitration Act 1996 or any statutory re-enactment or amendment thereof for the time being in force. The reference shall be conducted in accordance with the procedure set out in the Appendix to the Form of Tender or any amendment or modification thereof being in force at the time of the appointment of the arbitrator . Such arbitrator shall have full power to open up review and revise any decision opinion instruction direction certificate or valuation of the Employer's Representative or an adjudicator.

(b) Neither party shall be limited in the arbitration to the evidence or arguments put to the Employer's Representative or to any adjudicator pursuant to Clause 66(2) or 66(6) respectively.

(c) The award of the arbitrator shall be binding on all parties

(d) Unless the parties otherwise agree in writing any reference to arbitration may proceed notwithstanding that the Works are not then complete or alleged to be complete.

Witnesses

(12) (a) No decision opinion instruction direction certificate or valuation given by the Employer's Representative shall disqualify him from being called as a witness and giving evidence before a conciliator adjudicator or arbitrator on any matter whatsoever relevant to the dispute.

(b) All matters and information placed before a conciliator pursuant to a reference under sub-clause (5) of this Clause shall be deemed to be submitted to him without prejudice and the conciliator shall not be called as witness by the parties or anyone claiming through them in connection with any adjudication arbitration or other legal proceedings arising out of or connected with any matter so referred to him.

APPLICATION TO SCOTLAND ETC.

Application to Scotland

67 (1) If the Works are situated in Scotland (and unless the Contract otherwise provides) the Contract shall in all respects be construed and operate as a Scottish contract and shall be interpreted in accordance with Scots Law and the provisions of sub-clause (2) of this Clause shall apply.

(2) In the application of these Conditions and in particular Clause 66 thereof

(a) any reference to arbitration under these Conditions shall be conducted in accordance with the law of Scotland "The Scottish Arbitration Code 1999" prepared by the Scottish Council for International Arbitration the Chartered Institute of Arbitrators (Scottish Branch) and the Scottish Building Contract Committee together with the ICE Appendix (2001) thereto or any amendment to or modification of the Appendix being in force at the time of appointment of the arbitrator. Such arbitrator shall have full power to open up review and revise any decision opinion instruction direction certificate or valuation of the Employer's Representative or an adjudicator

(b) for any reference to the "Notice to Refer" there shall be substituted reference to the "Notice of Arbitration"

(c) the existing Clause 66(10) shall be deleted and replaced with new Clause 66(10) below.

" Appointment of arbitrator

66(10) (a) The arbitral tribunal shall be appointed by agreement of the parties.

President or Vice-President to act

(b) Failing agreement of the parties as aforesaid at sub-clause (a) above the following shall apply.

(i) Reference at Articles 3.5 3.6 3.7 3.8 and 6.2 of the Code to the Chairman of the Institute of Arbitrators (Scottish Branch) and to the Chairman of the Scottish Council for International Arbitration shall be deemed to be a reference to the President of the Institution of Civil Engineers as defined at (ii) below.

(ii) "President" means the President for the time being of the Institution of Civil Engineers or any Vice President acting on his behalf or such other person as may have been nominated in the arbitration agreement to appoint the arbitrator in default of agreement between the parties."

(d) the existing Clause 66(11) shall be deleted and replaced with new Clause 66(11) below.

" Arbitration – procedure and powers

66(11) (a) Neither party shall be limited in the arbitration to the evidence or arguments put to the Employer's Representative or to any adjudicator pursuant to Clause 66(2) or 66(6) respectively.

(b) Unless the parties otherwise agree in writing any reference to arbitration may proceed notwithstanding that the Works are not then complete or alleged to be complete. "

(e) notwithstanding any of the other provisions of these Conditions or of the Code (including in particular Articles 22.7 and 22.8) nothing therein shall exclude or be construed as excluding recourse to the Court of Session under Section 3 of the Administration of Justice of (Scotland) Act 1972

(f) where the Employer or the Contractor wishes to register the award of an arbitrator or the decision of an adjudicator in the Books of Council and Session for preservation and execution the other party shall on being requested to do so forthwith consent to such registration by subscribing the said decision before a witness and

(g) Delete existing Appendix to the Form of Tender Part 1 Item 22

(h) The application of Articles 1.6 1.8 1.9 and 1.10 of the Scottish Arbitration Code 1999 shall be subject to the provisions of Clause 66(2) of the Contract unless the Arbitrator otherwise directs.

Application to Northern Ireland

(3) If the Works are situated in Northern Ireland the Contract shall in all respects be construed and operate as a Northern Irish contract and shall be interpreted in accordance with the law of Northern Ireland.

Application elsewhere

(4) If the Works are situated in a country or jurisdiction other than England and Wales Scotland or Northern Ireland the Contract and the provisions for disputes settlement shall in all respects be construed and operate and be interpreted in accordance with the law of that country or jurisdiction.

NOTICES

Service of notices on Contractor

68 (1) Any notice to be given to the Contractor under the terms of the Contract shall be served in writing at the Contractor's principal place of business (or in the event of the Contractor being a Company to or at its registered office).

Service of notices on employer

(2) Any notice to be given to the Employer under the terms of the Contract shall be served in writing at the Employer's last known address (or in the event of the Employer being a Company to or at its registered office).

(3) Any notice to be given by or to the Employer's Representative under the terms of the Contract shall be delivered as the Employer's Representative may direct.

TAX MATTERS

Labour-tax fluctuations

69 (1) The rates and prices contained in the Contract shall be deemed to take account only of the levels and incidence in force at the date for return of tenders of

(a) the taxes levies contributions premiums or refunds (including national insurance contributions but excluding income tax and any levy payable under the Industrial Training Act 1982 or any statutory re-enactment or amendment thereof for the time being in force) which are by law payable by or to the Contractor and his sub-contractors in respect of their workpeople engaged on the Contract and

(b) any landfill tax payable by the Contractor or his sub-contractors pursuant to the Finance Act 1996 (Sections 39-71 and Schedule 5) and the Landfill Tax Regulations 1996 or any statutory re-enactment or amendment thereof for the time being in force

and shall not take account of any level or incidence of the aforesaid matters foreseeable or known to take effect at some later date.

(2) If after the date for return of tenders there shall occur any change in the level and/or incidence of any such taxes levies contributions premiums or refunds the Contractor shall so inform the Employer's Representative and the net increase or decrease shall be taken into account in arriving at the Contract Price. The Contractor shall supply the information necessary to support any consequent adjustment to the Contract Price. All certificates for payment issued after submission of such information shall take due account of the additions or deductions to which such information relates.

Value Added Tax

70 (1) The Contractor shall be deemed not to have allowed in his tender for the tax payable by him as a taxable person to the Commissioners of Customs and Excise being tax chargeable on any taxable supplies to the Employer which are to be made under the Contract.

Employer's Representative's certificates net of Value Added Tax

(2) All certificates issued by the Employer's Representative under Clause 60 shall be net of Value Added Tax.

In addition to the payments due under such certificates the Employer shall separately identify and pay to the Contractor any Value Added Tax properly chargeable by the Commissioners of Customs and Excise on the supply to the Employer of any goods and/or services by the Contractor under the Contract.

Disputes

(3) If any dispute difference or question arises between either the Employer or the Contractor and the Commissioners of Customs and Excise in relation to any tax chargeable or alleged to be chargeable in connection with the Contract or the Works each shall render to the other such support and assistance as may be necessary to resolve the dispute or question.

Clause 66 not applicable

(4) Clause 66 shall not apply to any dispute difference or question arising under this Clause.

THE CONSTRUCTION (DESIGN AND MANAGEMENT) REGULATIONS 1994

CDM Regulations 1994

71 (1) In this clause

(a) "the Regulations" means the Construction (Design and Management) Regulations 1994 or any statutory re-enactment or amendment thereof for the time being in force

(b) "Planning Supervisor" and "Principal Contractor" mean the persons so described in regulation 2(1) of the Regulations

(c) "Health and Safety Plan" means the plan prepared by virtue of regulation 15 of the Regulations.

(2) Where and to the extent that the Regulations apply to the Works and

(a) the Employer's Representative is appointed Planning Supervisor and/or

(b) the Contractor is appointed Principal Contractor

then in taking any action as such they shall state in writing that the action is being taken under the Regulations.

(3) (a) Any action under the Regulations taken by either the Planning Supervisor or the Principal Contractor and in particular any alteration or amendment to the Health and Safety Plan shall be deemed to be an instruction by the Employer's Representative or a notification by the Contractor as the case may be. Where appropriate additional payment and/or extension of time shall be assessed in accordance with Clause 13. Provided that the Contractor shall in no event be entitled to any additional payment and/or extension of time in respect of any such action to the extent that it results from any action lack of action or default on the part of the Contractor.

(b) If any such action of either the Planning Supervisor or the Principal Contractor could not in the Contractor's opinion reasonably have been foreseen by an experienced contractor the Contractor shall as early as practicable give written notice thereof to the Employer's Representative.

SPECIAL CONDITIONS

Special conditions

72 The following special Conditions form part of the Conditions of Contract.

(Note: Any special Conditions including contract price fluctuation which it is desired to incorporate in the Conditions of Contract should be numbered consecutively with the foregoing Conditions of Contract).

SHORT DESCRIPTION OF WORKS

All the design and Permanent and Temporary Works in connection with * ....

....

....

Form of Tender

(NOTE: The Appendix forms part of the Form of Tender)

To ....

....

....

GENTLEMEN,

Having examined the Employer's Requirements for the above mentioned Works as set out in your communication of...................the Conditions of Contract (and the matters set out in the Appendix hereto) we offer to design construct and complete the said Works in conformity with the said Requirements and Conditions of Contract and with the other documents listed below for the sum of

£..

....(£..)
or such other sum as may be ascertained in accordance with the said Conditions of Contract.

If our tender is accepted, we will if required provide security for the due performance of the Contract as stipulated in the Conditions of Contract and the Appendix hereto.

Unless and until a formal Agreement is prepared and executed this tender together with your written acceptance thereof shall form a binding Contract between us.

We understand that you are not bound to accept the lowest or any tender you may receive.

We are, Gentlemen,

Yours faithfully

Signature ....

Address ....

....

Date.......................................

Other documents referred to above.
Appendix to the Form of Tender Parts 1 & 2.

....

....

....

....
* Complete as appropriate

FORM OF TENDER (APPENDIX)

(*NOTE: Relevant Clause numbers are shown in brackets*)

Appendix - Part 1 (*to be completed prior to the invitation to tender*)

1. Name of the Employer(Clause 1(1)(a)) ....

   Address..

   ....

2. Name of Employer's Representative (*if known*)(Clauses 1(1)(c) and 2(2))..

   ....

3. Defects Correction Period (Clause 1(1)(p)) .... weeks

4. Parts or Sections of the Works or design which shall not be sub-contracted without the Employer's Representative's prior approval (Clause 4(2))

....

....

....

5. Quality Assurance (Clause 8(3)) Required/not required

6 Form of Agreement (Clause 9) Required/not required

   If required Signed/Sealed

7. Performance Bond (Clause 10(1)) Required/not required

   Amount if required .... % of estimated Contract Price

8. Minimum amount of Third Party Insurance (persons and property) (Clause 23(3)) £
   for each and every occurrence.

9. Commencement Date (*if known*) (Clause 41(1)(a)) ....

10. Time for Completion (Clause 43) [a]

    EITHER for the whole of the Works .... weeks

    OR for Sections of the Works (Clause 1(1)(r)) [b]

    Section A .... weeks

    Section B .... weeks

    Section C .... weeks

    the remainder of the Works .... weeks

11. Liquidated Damages for Delay (Clause 47)

	per day/week	limit of liability [c]
EITHER for the whole Works	£..	£..
OR for Section A (*as above*)	£..	£..
Section B (as above)	£..	£..
Section C (*as above*)	£..	£..
the remainder of the Works (*as above*)	£..	£..

12. Vesting of Materials not on Site (Clause 54(4) & 60(1)(d)) (*if required by the Employer*) [d]

1 .... 4...

2 .... 5...

3 .... 6...

13. Maximum percentage of the value of goods and materials to be included in interim certificates (Clause 60(2)(b)(ii)) .... %

14. Minimum amount of interim certificates (Clause 60(3)) £..

15. Rate of retention. (Clause 60(5)(a)) %

16. Limit of retention (if any) £..

17. Bank whose Base Lending Rate is to be used (Clause 60(7) ....

18. Name of the Planning Supervisor (Clause 71(1)(b))

....

Address ....

19. Name of the Principal Contractor (*if appointed*) (Clause 71(1)(b))

....

Address ....

20. The arbitration procedure to be used is (Clause 66(11)(a)

(a) The Institution of Civil Engineers' Arbitration Procedure (1997) [e]

or

(b) The Construction Industry Model Arbitration Rules [e]

If no deletion is made The Institution of Civil Engineer's Arbitration Procedure (1997) shall be used.

[a] If not stated, it is to be completed by the Contractor in Part 2 of the Appendix.

[b] To be completed if required, with brief description. Where Sectional completion applies the item for "the Remainder of the Works" must be used to cover the balance of the Works if the Sections described do not in total comprise the whole of the Works.

[c] Delete where not required.

[d] (If used) Materials to which the Clauses apply must be listed in Part 1 (Employer's option) or Part 2 (Contractor's option)

[e] Delete one as appropriate.

Appendix - Part 2.

(To be completed by the tenderer)

1. Insurance Policy excesses (Clause 25(2))

   Insurance of the Works (Clause 21(1)) £...

   Third Party (property damage) (Clause 23(1)) £

2. Time for Completion of the Works (Clause 43) *(if not completed in Part 1 of the Appendix)*

   EITHER for the whole of the Works ....weeks

   OR for Sections of the Works (Clause 1(1)(r))

   Section A .... weeks

   Section B .... weeks

   Section C .... weeks

   the remainder of the Works .... weeks

3. Vesting of Materials not on Site (Clause 54(4) & 60(1)(c))
   (at the option of the Contractor – see [d] in Part 1)

   1 .... 4. ....

   2 .... 5. ....

   3 .... 6. ....

4. Percentage adjustment for Prime Cost Items (Clause 58(2)) ....%

5. Name of Contractor's Representative..

6. Contractor's Designer (Clause 4(2)(a))

   The design work for which the Contractor is responsible will be carried out by

   ....

   ....

   ....

   ....

   ....

Elements for which the Contractor has not yet appointed a designer.

....

....

....

ICE FORM OF DEFAULT BOND

Date

Parties

[Names addresses and Company Numbers if applicable]

SURETY (1)

CONTRACTOR (2)

EMPLOYER (3)

Background

(A) By a Contract defined in the Schedule hereto the Contractor has agreed with the Employer to design construct and complete the Works.

(B) The Surety has agreed to provide this Bond in favour of the Employer in order to guarantee the performance by the Contractor of his obligations under the Contract.

Surety's obligation **1**

(1) If the Contractor fails to pay the Excess Sum within 28 days of receipt by the Contractor of a copy of a certificate issued by the Employer's Representative under clause 65(5) of the Contract the Surety hereby guarantees to the Employer that the Surety shall subject to the terms and conditions of this Bond pay the Excess Sum in accordance with Clause 1(3) up to the Bond Amount.

(2) It shall be a condition precedent to payment by the Surety that the Employer serve on the Surety a copy of the certificate issued by the Employer's Representative under clause 65(5) of the Contract as served on the Contractor and certified by the Employer's Representative as being a true copy of such certificate.

(3) Subject to Clause 1(4) payment by the Surety shall be made not later than 14 days after the later of

(a) the expiry of the 28 day period referred to in Clause 1(1) (save in respect of any payment made by the Contractor within that time) and

(b) service on the Surety of the copy certificate referred to in Clause 1(2).

(4) If the Surety objects to the contents of or entitlement to issue a certificate under clause 65(5) of the Contract in respect of which the Employer seeks payment from him the Surety shall have the right to refer the matter to adjudication in accordance with the adjudication provisions contained in sub-clauses 66(6) to 66(8) of the Contract as if the Surety were a party to the Contract in place of the Contractor.

(5) Any adjudication under Clause 1(4) shall be commenced by the Surety within 14 days of receipt by the Surety of the documents referred to in Clause 1(2) and the Surety shall have no right to refer the matter to adjudication after that time.

(6) If the content of or entitlement to issue the Certificate under clause 65(5) of the Contract in respect of which payment is sought by the Employer is or has been the subject of an adjudication between the Employer and Contractor under the Contract (in respect of which both parties have made submissions to the adjudicator) the Surety agrees to be bound by the result of such adjudication and shall have no right to refer the matter to adjudication under Clause 1(4).

(7) In the case of an adjudication under Clause 1(4) payment by the Surety shall be made within 7 days of the decision in such adjudication.

Surety's rights **2** (1) The Surety shall be entitled to receive copies of any notice given by the Employer under clause 65(1) of the Contract (with any Employer's Representative's certificate referred to) within 7 days of such notice or certificate being served on the Contractor.

(2) The Surety shall be entitled at any time within 7 days of receipt by the Surety of the copy certificate referred to in Clause 1(2)

(a) to request the Employer to provide the Surety with such further information and documentation as the Surety may reasonably require to verify the Excess Sum (including information or documentation held by the Employer's Representative) and/or

(b) to request to inspect the Site and the Works upon reasonable notice (the Employer may require that a representative of the Employer accompanies the Surety during such inspections).

Accounting **3** (1) If the Excess Sum is subsequently determined by reason of a subsequent certificate issued by the Employer's Representative under clause 65(5) of the Contract or by adjudication arbitration litigation or agreement between the Surety and the Employer to be less than the amount paid by the Surety the difference (if the Excess Sum has already been paid by the Surety) shall be repaid by the Employer to the Surety with Interest within 14 days (or such other period as the adjudicator arbitrator or Court may direct) after the date of such determination or agreement.

(2) If the Excess Sum is subsequently determined or agreed to be greater than the amount already paid by the Surety any difference (up to the Bond Amount) shall be paid by the Surety to the Employer with Interest within 14 days (or such other period as the adjudicator arbitrator or Court may direct) after the date of such determination or agreement.

Interest **4** Subject to the amount payable by the Surety being varied in accordance with Clause 3 above in the event that any amount payable by either the Surety or the Employer under this Bond is not made by the date determined by Clause 1(3) or in accordance with Clause 3 (the Due Date) then the payer shall pay Interest on the sum from the Due Date until the date of payment.

Expiry	**5**	Save in respect of any failure to pay the Excess Sum in respect of which a claim in writing has been received beforehand from the Employer by the Surety this Bond shall expire on the later of the date stated in a certificate of substantial completion issued by the Employer's Representative and the Final Expiry Date.
Forbearance	**6**	The Surety shall not be discharged or released by any alteration variation or waiver of any of the terms and conditions and provisions of the Contract or in any extent or nature of the Works and no allowance of time by the Employer under or in connection with the Contract or the Works shall in any way release reduce or affect the liability of the Surety under this Bond.
Governing law	**7**	This Bond shall be governed and construed in accordance with the laws of the country named in the Schedule ("the Country") and the courts of the Country shall have exclusive jurisdiction.
Assignment	**8**	This Bond may only be assigned by the Employer with the prior consent of the Surety and the Contractor which consent shall not be unreasonably withheld. In the event of any such assignment the Employer and assignee shall remain jointly and severally liable for any repayment due to the Surety under Clause 3(1). Notice of any assignment shall be given to the Surety as soon as practicable.

EXECUTED AS A DEED

on behalf of the Surety	(1)	director	____________________
	(2)	director/secretary	____________________
on behalf of the Contractor	(1)	director	____________________
	(2)	director/secretary	____________________
on behalf of the Employer	(1)	director	____________________
	(2)	director/secretary	____________________

SCHEDULE

Address for Service	**Contractor:**
	Tel: Fax:
	Employer:
	Tel: Fax:
	Surety:
	Tel: Fax:
"Bond Amount"	means the sum of £[ ] ([ ] pounds) being the maximum aggregate liability of the Surety under this Bond.
"Contract"	means the contract [made between the Employer and the Contractor dated the [ ] day of [ ] [ ]] / [to be entered into between the Employer and the Contractor] incorporating the ICE Conditions of Contract 7th Edition.
"Excess Sum"	an amount certified as due to the Employer under Clause 65(5) of the Contract.
"Employer's Representative" "Works" and "Site"	have the same meaning as in the Contract.
"Final Expiry Date"	means the [ ] day of [ ] [ ].
"Interest"	means the rate of interest specified in the Contract.
"The Country"	means [England and Wales] [Scotland] [Northern Ireland]

(This Bond has been drafted in collaboration with S. J. BERWIN & Co)

The Institution of Civil Engineers | Association of Consulting Engineers | Civil Engineering Contractors Association

This clause has been prepared by The Institution of Civil Engineers, the Association of Consulting Engineers and the Civil Engineering Contractors Association, in consultation with the Government in its revised form, for use in appropriate cases as a Special Condition of the Conditions of Contract for use in connection with the ICE Conditions of Contract SEVENTH EDITION dated September 1999.

CONTRACT PRICE FLUCTUATIONS CIVIL ENGINEERING WORK

(1) The amount payable by the Employer to the Contractor upon the issue by the Employer's Representative of an interim certificate pursuant to Clause 60(2) or of the final certificate pursuant to Clause 60(4) (other than amounts due under this Clause) shall be increased or decreased in accordance with the provisions of this Clause if there shall be any changes in the following Index Figures compiled by the Department of the Environment Transport and the Regions and published by The Stationery Office in the Monthly Bulletin of Indices "1990 Series Civil Engineering Formula Indices" and "1990 Series Structural Steelwork Formula Indices"

(a) the Index of Labour and Supervision (Index 1)

(b) the Index of providing and maintaining Contractor's Equipment (Index 2 - Plant and Road Vehicles)

(c) the Indices of material prices applicable to those materials listed in sub-clause (4) of this Clause.

The net total of such increases and decreases shall be given effect to in determining the Contract Price.

(2) For the purpose of this Clause

(a) 'Final Index Figure' shall mean any Index Figure appropriate to sub-clause (1) of this Clause not qualified in the said bulletin as provisional

(b) 'Base Index Figure' shall mean the appropriate Final Index Figure applicable to the date 42 days prior to the date for the return of tenders

(c) 'Current Index Figure' shall mean the appropriate Final Index Figure to be applied in respect of any certificate issued or due to be issued by the Employer's Representative pursuant to Clause 60 and shall be the appropriate Final Index Figure applicable to the date 42 days prior to

(i) the due date (or extended date) for completion or

(ii) the date certified pursuant to Clause 48 of completion of the whole of the Works or

(iii) the last day of the period to which the certificate relates

whichever is the earliest.

Provided that in respect of any work the value of which is included in any such certificate and which work forms part of a Section for which the due date (or extended date) for completion or the date certified pursuant to Clause 48 of completion of such Section precedes the last day of the period to which the certificate relates the Current Index Figure shall be the Final Index Figure applicable to the date 42 days prior to whichever of these dates is the earliest.

(d) The 'Effective Value' in respect of the whole or any Section of the Works shall be the difference between

(i) the amount which in the opinion of the Employer's Representative is due to the Contractor under Clause 60(2) (before deducting retention) or the amount due to the Contractor under Clause 60(4) (but in each case before deducting sums previously paid on account) less any amounts for Dayworks Nominated Sub-contractors or any other items based on actual cost or current prices and any sums for increases or decreases in the Contract Price under this Clause and

(ii) the amount calculated in accordance with (i) above and included in the last preceding interim certificate issued by the Employer's Representative in accordance with Clause 60.

Provided that in the case of the first certificate the Effective Value shall be the amount calculated in accordance with sub-paragraph (i) above.

(3) The increase or decrease in the amounts otherwise payable under Clause 60 pursuant to sub-clause (1) of this Clause shall be calculated by multiplying the Effective Value by a Price Fluctuation Factor which shall be the net sum of the products obtained by multiplying each of the proportions given in (a) (b) and (c) of sub-clause (4) of this Clause by a fraction the numerator of which is the relevant Current Index Figure minus the relevant Base Index Figure and the denominator of which is the relevant Base Index Figure.

(4) For the purpose of calculating the Price Fluctuation Factor the proportions referred to in sub-clause (3) of this Clause shall (irrespective of the actual constituents of the work) be as follows and the total of such proportions shall amount to unity

(a) 0._____* in respect of labour and supervision costs subject to adjustment by reference to the Index referred to in sub-clause (1)(a) of this Clause. (Index 1)

(b) 0._____* in respect of costs of provision and use of Contractor's Equipment which shall be subject to adjustment by reference to the Index referred to in sub-clause (1)(b) of this Clause. (Index 2)

(c) the following proportions in respect of the materials named which shall be subject to adjustment by reference to the relevant indices referred to in sub-clause (1)(c) of this Clause

0.___ *	in respect of Aggregates. (Index 3)
0.___ *	in respect of Bricks and Clay Products. (Index 4)
0.___ *	in respect of Cements. (Index 5)
0.___ *	in respect of Ready Mixed Concrete. (Index 6)
0.___ *	in respect of Cast and Spun Iron Products. (Index 7)
0.___ *	in respect of Plastics Products. (Index 8)
0.___ *	in respect of Coated Macadam and Bituminous Products. (Index 9)
0.___ *	in respect of Fuel for plant to which the DERV Fuel Index will be applied. (Index 10)
0.___ *	in respect of Fuel for plant to which the Gas Oil Fuel Index will be applied. (Index 11)
0.___ *	in respect of Timber. (Index 12)
0.___ *	in respect of Steel for Reinforcement. (Index 13)
0.___ *	in respect of Metal Sections. (Index 14)
0.___ *	in respect of Sheet Steel Piling. (Index 15)
0.___ *	in respect of Structural Steelwork Materials for Civil Engineering Works (Index S3)

(d) 0. 10 in respect of all other costs which shall not be subject to any adjustment

Total 1.00

(5) Provisional Index Figures in the Bulletin referred to in sub-clause (1) of this Clause may be used for the provisional adjustment of interim valuations but such adjustments shall be subsequently recalculated on the basis of the corresponding Final Index Figures.

(6) Clause 69 – Tax Fluctuations – shall not apply except to the extent that any matter dealt with therein is not covered by the Index of the Cost of Labour in Civil Engineering Construction.

* To be filled in by the Employer prior to inviting tenders.

The Institution of Civil Engineers	Association of Consulting Engineers	Civil Engineering Contractors Association

This clause has been prepared by The Institution of Civil Engineers, the Association of Consulting Engineers and the Civil Engineering Contractors Association, in consultation with the Government in its revised form, for use in appropriate cases as a Special Condition of the Conditions of Contract for use in connection with the ICE Conditions of Contract SEVENTH EDITION dated September 1999.

CONTRACT PRICE FLUCTUATIONS STRUCTURAL STEELWORK

(1) The amount payable by the Employer to the Contractor upon the issue by the Employer's Representative of an interim certificate pursuant to Clause 60(2) or of the final certificate pursuant to Clause 60(4) (other than amounts due under this Clause) shall be increased or decreased in accordance with the provisions of this Clause if there shall be any changes in the following Index Figures compiled by the Department of the Environment Transport and the Regions and published by the Stationery Office in the Monthly Bulletin of Indices "1990 Series Structural Steelwork Formula Indices"

(a) the Index of Labour in fabrication of Structural Steelwork and erection of Structural Steelwork (Index S1)

(b) the Index for Structural Steelwork Materials (Civil Engineering Works) (Index S3).

The net total of such increases and decreases shall be given effect to in determining the Contract Price.

(2) For the purpose of this Clause

(a) 'Structural Steelwork' shall mean those items of work listed in sub-clause (6) of this Clause and shall include any alterations as may be ordered under Clause 51 involving work of a description which in the opinion of the Employer's Representative is similar to the description of the items so listed

(b) 'Final Index Figure' shall mean any Index Figure appropriate to sub-clause (1) of this Clause not qualified in the said bulletin as provisional

(c) 'Base Index Figure' shall mean the appropriate Final Index Figure applicable to the date 42 days prior to the date for the return of tenders

(d) 'Current Index Figure' shall mean the appropriate Final Index Figure to be applied in respect of any certificate issued or due to be issued by the Employer's Representative pursuant to Clause 60 being such figure applicable

(i) in respect of labour employed in fabrication - to a date 56 days prior to the last day of the period to which the certificate relates or

(ii) in respect of labour employed in erection - to a date 14 days prior to the last day of the period to which the certificate relates or

(iii) in respect of materials specifically purchased for inclusion in the Works - to the date of delivery to the fabricator's premises (of which date the Contractor shall produce such evidence relating to gross tonnages delivered as the Employer's Representative may reasonably require) or

(iv) in respect of materials (if any) not specifically purchased for inclusion in the Works – to the date of the last of the deliveries referred to in sub-paragraph (iii) of this paragraph

as the case may be.

Provided always that should the due date (or extended date) for completion or the date certified pursuant to Clause 48 of completion of the whole of the Works precede any of the aforesaid then such due date or extended date or certified date whichever is earliest shall be substituted for those aforesaid.

Provided further that if in respect of any work which forms part of a Section and whose value is included in any certificate the due date (or extended date) for completion of that Section or the date certified pursuant to Clause 48 of completion of that Section precede any of the dates aforesaid in sub-paragraphs (i) to (iv) above then such due date or extended date or certified date whichever is the earliest in respect of that Section shall be substituted for those aforesaid.

(e) The 'Effective Value' in respect of the whole or any Section of the Works shall be the difference between

(i) the amount which in the opinion of the Employer's Representative is due to the Contractor under Clause 60(2) (before deducting retention) or the amount due to the Contractor under Clause 60(4) (but in each case before deducting sums previously paid on account) less any amounts for Dayworks Nominated Sub-contractors or any other items based on actual cost or current prices and any sums for increases or decreases in the Contract Price under this Clause and

(ii) the amount calculated in accordance with (i) above and included in the last preceding interim certificate issued by the Employer's Representative in accordance with Clause 60.

Provided that in the case of the first certificate the Effective Value shall be the amount calculated in accordance with sub-paragraph (i) above.

(3) The Effective Value shall be apportioned between labour and materials in the following manner that is to say

(a) labour

(i) employed in erection - by multiplying the total tonnage erected during the period to which the certificate relates by the average cost per tonne entered at (b) of sub-clause (5) of this Clause

(ii) employed in fabrication and delivery - by deducting the summation of the values calculated in respect of materials and in respect of labour employed in erection from the Effective Value

(b) materials

by multiplying the total tonnage of steel delivered to the Site for inclusion in the Works during the period to which the certificate relates by the average price per tonne entered at (a) of sub-clause (5) of this Clause.

(4) (a) The increase or decrease in the amounts otherwise payable under Clause 60 pursuant to sub-clause (1) of this Clause shall be calculated by multiplying each portion of the Effective Value by a fraction the numerator of which is the product of 0.90 and the difference between the relevant Current Index Figure and the relevant Base Index Figure and the denominator of which is the relevant Base Index Figure.

(b) The relevant indices to be used in connection with this sub-clause are

(i) for labour in fabrication and erection-the Index referred to in sub-clause (1)(a) of this Clause (Index S1)

(ii) for materials – the Index referred to in sub-clause (1)(b) of this Clause (Index S3).

(5) For the purpose of the apportionment in sub-clause (3) of this Clause the full average costs per tonne (inclusive of all associated labour plant power maintenance overheads and profit) to be used are

(a) Materials delivered to fabricators premises £ per tonne *

(b) Erection £ per tonne *

(c) Subject to sub-paragraphs (i) (ii) and (iii) of this paragraph the relevant figures to be used in connection with this sub-clause shall be in accordance with sub-clause (2) of this Clause.

Provided that in respect of materials

(i) the Current Index Figure subsequent to the first established Current Index Figure pursuant to sub-clause (2) (d) (iii) of this Clause shall not be used until a tonnage of steel greater than the tonnage to which the Current Index Figure first established applies has been delivered to the Site for inclusion in the Works and

(ii) for the purpose of establishing the appropriate subsequent Current Index Figures to apply to all later deliveries of steel to the Site for inclusion in the Works the provisions of sub-paragraph (i) of this paragraph shall apply mutatis mutandis and

(iii) the Current Index Figure referred to in sub-clause (2) (d) (iv) of this Clause shall not be used until the total tonnage of steel delivered to the Site for inclusion in the Works exceeds the total tonnage of steel specifically purchased for inclusion in the Works and delivered to the fabricator's premises.

(6) For the purposes of this Clause the expression 'Structural Steelwork' shall comprise only those items listed hereunder †

Bill No †	Page No †	Item No †

(7) Provisional Index Figures in the Bulletin referred to in sub-clause (1) of this Clause may be used for the provisional adjustment of interim valuations but such adjustments shall be subsequently recalculated on the basis of the corresponding Final Index Figures.

(8) Clause 69 – Tax Fluctuations – shall not apply except to the extent that any matter dealt with therein is not covered by the Index of the Cost of Labour in fabrication of steelwork and erection of steelwork.

* To be filled in by the Contractor at the time of tendering.
† To be filled in by the Employer prior to inviting tenders.

The Institution of Civil Engineers | Association of Consulting Engineers | Civil Engineering Contractors Association

This clause has been prepared by The Institution of Civil Engineers, the Association of Consulting Engineers and the Civil Engineering Contractors Association, in consultation with the Government in its revised form, for use in appropriate cases as a Special Condition of the Conditions of Contract for use in connection with the ICE Conditions of Contract SEVENTH EDITION dated September 1999.

CONTRACT PRICE FLUCTUATIONS CIVIL ENGINEERING WORK AND STRUCTURAL STEELWORK

(1) This Clause shall apply only to those contracts which incorporate the Contract Price Fluctuation Clause Revised February 1999 for Civil Engineering work and the Structural Steelwork Clause Revised February 1999 (referred to in this Clause as the CEW Clause and the SS Clause respectively).

(2) For the purposes of this Clause

(a) 'Civil Engineering work' shall mean all Works with the exception of Structural Steelwork and

(b) ' Structural Steelwork' shall mean the work defined in sub-clause (2)(a) of the SS Clause.

(3) The Effective Value (as defined both in the CEW Clause and in the SS Clause) shall be sub-divided to show the amounts included in respect of the Civil Engineering work and the Structural Steelwork. The amount in respect of the former shall then be treated as if it were the Effective Value as defined in the CE Clause and adjusted in accordance with the provisions of that Clause. The amount in respect of the latter shall then be treated as if it were the Effective Value as defined in the SS Clause and adjusted in accordance with that Clause.

I.C.E. CONDITIONS OF CONTRACT STANDING JOINT COMMITTEE (CCSJC)

AMENDMENT

to take into account

THE FINANCE ACT 2001

The Sponsoring Bodies of the Conditions of Contract and Forms of Tender, Agreement, Bond, and Contract Schedule for use in connection with Works of Civil Engineering Construction have approved amendments to cover the introduction of the Finance Act 2001.

The new Clause 69(1)(c) makes it clear that only those rates of Aggregate Levy chargeable under the above Act and in force at the date for return of tenders shall be deemed to have been included in the tender rates, whether or not future increases or decreases are then foreseeable. Increases or decreases after the date for return of tenders are covered in Clause 69(2). It is recommended that this revised clause be incorporated into contracts agreed forthwith.

The Conditions of Contract with amendments that are notified are as follows:-

ICE 7th Edition (September 1999)

- (Reference ICE/ScotArb/April 2001).

ICE 6th Edition (January 1991), reprinted 1995, 1997 and 1998

- Guidance Note (March 1995)
- (Reference ICE/6th Edition/Tax/February 1998)
- (Reference ICE/6th Edition/HGCR/March 1998)
- (Reference ICE/6th Edition/Pub/September 1998)
- (Reference ICE/CPF/February 1999)
- (Reference ICE/3rd Parties/December 1999)
- (Reference ICE/ScotArb/April 2001).

ICE 5th Edition (June 1973) (Revised January 1979) (Reprinted with amendments January 1986)

- Guidance Note 8
- Guidance Note (March 1995)
- Reference ICE/5th Edition/Tax/February 1998
- Reference ICE/5th Edition/HGCR/March 1998
- Reference ICE/CPF/February 1999
- Reference ICE/3rd Parties/December 1999
- Reference ICE/ScotArb/April 2001

Design and Construct 2nd Edition (September 2001)

Design and Construct 1st Edition (1992), reprinted 1997

- Guidance Note (March 1995)
- Reference ICE/D&C/Tax/February 1998
- Reference ICE/D&C/HGCR/March 1998
- Reference ICE/D&C/Pub/September 1998
- Reference ICE/CPF/February 1999
- Reference ICE/3rd Parties/December 1999
- Reference ICE/ScotArb/April 2001

Ground Investigation First Edition (October 1983)

- Guidance Note (March 1995)
- Reference ICE/GI/HGCR/October 1998
- Reference ICE/3rd Parties/December 1999
- Reference ICE/ScotArb/April 2001

Clause 69 **TAX MATTERS**

Add sub-clause 69(1)(c) below:-

" (c) any aggregate levy payable by the Contractor or his sub-contractors pursuant to the Finance Act 2001 or any statutory re-enactment or amendment thereof for the time being in force. "

CCSJC OPINIONS

The ICE Conditions of Contract Standing Joint Committee is constituted of members appointed by the sponsors of the ICE Conditions of Contract: The Institution of Civil Engineers, The Association of Consulting Engineers and The Civil Engineering Contractors Association, and its function is to keep the use of the above Forms of Contract under review. The Terms of Reference of the CCSJC exclude the provision of any legal interpretation.

Tendering for Civil Engineering Contracts

recommended for use with the

ICE Conditions of Contract

In the United Kingdom

Tendering for Civil Engineering Contracts

recommended for use with the

ICE Conditions of Contract

in the United Kingdom

The Institution of Civil Engineers
Association of Consulting Engineers
Civil Engineering Contractors Association

Thomas Telford

Published by Thomas Telford Publishing, Thomas Telford Limited, 1 Heron Quay, London E14 4JD on behalf of:

The Chief Executive and Secretary
The Institution of Civil Engineers
One Great George Street
London SW1P 3AA

The Chief Executive
Association of Consulting Engineers
Alliance House
12 Caxton Street
London SWIH 0QL

The Director
Civil Engineering Contractors Association
Construction House
56-64 Leonard Street
London EC2A 4JX

'Tendering for Civil Engineering Contracts' recommended for use with the ICE Conditions of Contract in the United Kingdom is produced by the ICE Conditions of Contract Standing Joint Committee (CCSJC). It updates and replaces the 'Guidance on the Preparation, Submission and Consideration of Tenders for Engineering Contracts' recommended for use in the United Kingdom which was produced by the same organisation in 1983.

The CCSJC is a permanent joint committee which has prepared and will keep under review the use of this document and will consider any suggestions for amendment, which should be addressed to The CCSJC c/o Chief Executive and Secretary, The Institution of Civil Engineers, One Great George Street, London SWIP 3AA. Revision to the document will be made when such action seems warranted.

9 8 7 6 5 4 3 2 1

ISBN 0 7277 2853 9

A catalogue record for this book is available from the British Library

Printed and bound in Great Britain

CONTENTS

FOREWORD

This publication has been prepared by the Conditions of Contract Standing Joint Committee (CCSJC) specifically to assist users of ICE Conditions of Contract with the procedures between the start of the tender process and the award of the contract. It does not purport to provide legal interpretation but does represent the view of the CCSJC on what constitutes good practice in the conduct of civil engineering projects.

The Sponsoring Bodies of the CCSJC are The Institution of Civil Engineers (ICE), the Association of Consulting Engineers (ACE) and the Civil Engineering Contractors Association (CECA). Membership of the CCSJC will comprise:

- An independent Chairman appointed by the ICE with the agreement of the Sponsors;
- Three members appointed by the ICE to present the views of the clients;
- Three members appointed by the ACE to present the views of the consultants;
- Three members appointed by the CECA to present the views of all contractors;
- One Fellow or Member of the ICE to provide professional advice of a practical nature including legal advice.

1 INTRODUCTION

1.1 **1** The ICE Conditions of Contract are widely used as the legal framework for construction contracts. These deal with the administration of the Contract and the key role of the Engineer who acts in a professional capacity for the Employer as an interface with the Contractor.

2 The various documents on which the tender invitation is based (which then comprise the Contract) are prepared and issued by the Engineer. These documents are Conditions of Contract, Form of Tender, Drawings, Specification, Bill of Quantities (or pricing schedule) and Form of Agreement together with any other agreed document or information.

1.2 **1** This publication sets out the procedures which should be adopted for the selection of tenderers, the tender process, dealing with queries during the tender, tender receipt and adjudication and necessary actions prior to the award of a contract. The document represents 'Good Practice'. Its content is not mandatory nor can the procedures set out be considered to be enforceable.

2 Careful planning and management of the tender process is essential to ensure fair competition. The document has been drafted bearing in mind the benefits of team working and current procurement initiatives.

1.3 The section on 'Tender Documentation and Procedure' of the ICE Conditions of Contract, Measurement Version, 7th Edition Guidance Notes makes specific reference to this publication. Particular matters in regard to 'Tender Procedures' for contracts governed by the ICE Design and Construct Conditions of Contract are set out in the Guidance Notes to those Conditions published in 1993.

1.4 Reference should be made to the latest version of the FIDIC publication 'Tendering Procedure' for advice on contract practice overseas.

1.5 Where UK Contracts Regulations and/or EC Public Procurement Directives[1] are applicable to procedures for the advertising and letting of contracts for public works or works under government licence or other projects, the particular requirements set out in such legislation will need to be followed. Also many large Employers and/or Funders have rules which relate to procurement procedures for advertising, consideration, reasons for rejection or acceptance and letting of contracts for civil engineering works.

1.6 The definition of terms, set out in whichever standard Conditions of Contract are adopted, should be preserved and used in all documents associated with the tender.

1.7 The tender process is integral with the activities of design and the preparation of drawings and specifications. It is important that the tender process is co-ordinated with other activities, and planned and programmed into the overall project plan. The Client should name an individual as the tender manager who is responsible for the process of document preparation, integration and the monitoring of stages leading to the award of a contract.

[1] See Section 8, Bibliography, for references to Statutory Instruments and EEC Directives.

2 GENERAL

Open and Selective Tendering **2.1**

1 Open tendering, i.e. allowing bids from any contractor interested, may be undesirable because it involves too many contractors in abortive tendering that wastes resources and is inefficient. Selective tendering, based on approved lists or pre-qualification to produce a short list, is recommended as being the better procedure, affording maximum efficiency and economic advantage all round. If open tendering is to be adopted, tenderers should be informed at the time of issuing the tender documents of the number of contractors to whom issue has been made.

2 There may be over-riding requirements under UK Contracts Regulations or EC Public Procurement Directives which modify the foregoing. Such legislation includes directives relating to public procurement.

3 For some contracts, two stage tendering may be implemented. This involves a selection based on an indicative price with a firm price being agreed as the work content is finalised. In these special circumstances consideration may be given to making some payment to tenderers for their part in the development of the design. This is a different basis for arriving at a contract from that envisaged in this document, and it can be expensive and time consuming. There is a wide divergence of opinion on two stage tendering. This form of tendering is generally only considered to be appropriate where it is desired to select a preferred tenderer before the design is fully developed. It should not be used as a general procedure for Works which have been designed sufficiently for pricing. However, for large design and build projects which would benefit from early contractor input and where the preparation of a full tender submission is complex and involves large teams of specialists, it may have benefits.

Pre-qualification of Contractors **2.2**

1 With selective tendering, pre-qualification of contractors is normally required to assist in compiling a list of firms most suited to receive invitations to tender and which can undertake the Works. (Reference is made in paragraph 2.3.3 to the UK qualification services provided by Constructionline). Contractors invited to pre-qualify should be asked to submit details of their experience relevant to the particular type of work in the location or circumstances applying. The amount of information requested should reflect the technical content of the works in question and their value. The factors considered should be assessed under the following four headings:

(a) *The contractor's financial standing*. Whether the Contractor is financially stable and/or has the guaranteed backing of a larger group to withstand any financial problems that may occur during the contract. Such assessment normally includes examination of accounts, annual reports (if a public company) or a confidential report from the firm's bank.

(b) *Technical and organisational ability*. Whether the Contractor has adequate capacity and ability to undertake the Works to the required quality at the time in question and whether suitably experienced staff are available. Enquiries made of other Employers and Engineers may assist in assessing this aspect of the firm based on past experience.

(c) *General experience and performance record*. Whether the Contractor has had sufficient experience in the particular type and magnitude of works and has a satisfactory performance reputation. This knowledge is best acquired by interviewing the contractor rather than relying entirely on published documents or the views of others.

(d) *Health and Safety*. Whether the firm has a Health and Safety Policy, and how it implements the Construction (Design and Management) Regulations 1994. Other questions relating to the Contractor's capacity to act as Principal Contractor, its record on Health & Safety issues, general policy, organisation, planning and monitoring of H & S are appropriate.

2 The extent to which contractors should be asked to pre-qualify should depend on the nature of the work and the knowledge that is already available to the Engineer/Employer. Contractors whose qualifications and past performance records are already well known should not be required to pre-qualify on every occasion,

Preferred Contractors Lists 2.3

1 There is a case for Employers maintaining standing lists of preferred Contractors (for instance under a partnering framework) or for approved Contractors, as opposed to compiling selected lists on an ad-hoc basis. For each project the list is dependent on the size, nature and continuity of the Employer's programme of works. Regardless of whether ad-hoc approved lists are compiled following public advertisement, or otherwise exist as standing lists, the method should aim to identify only those firms who have the necessary technical and financial resources to complete the contract satisfactorily.

2 Standing lists should be stratified by the types and categories of work and by value against which the qualifications of contractors are approved, and it is important that the factors on which entry is based are made known. It is also important that standing lists should not be allowed to stagnate by default with no incentive for those included to maintain the necessary standard. Individual performance should be reviewed at the end of each contract and the opportunity taken to transfer the contractor to another category if appropriate: the claims of new contractors for inclusion should be reviewed at least annually.

3 Another means which may assist in compiling lists is the UK qualification service provided by Constructionline. Sponsored by the Department of the Environment, Transport and the Regions it provides registers of qualified works contractors after carrying out checks on their capabilities, technical competence, managerial standards and financial strength. The information is recorded on a database which clients can access and use as a means of identifying contractors with the required skills. Use of this register may also help avoid unnecessary costs otherwise incurred by clients and contractors in duplicating qualification procedures.

Number of Tenderers Selected 2.4

1 The objective of selective tendering is to limit the number of contractors tendering to a sensible level. It is generally accepted as good practice that the number of contractors invited to tender should not be less than four nor more than eight. If significant or complicated design work is included as part of the Works it is recommended that only three firms or consortia should be invited to tender and two stage tendering may then be appropriate. The report of CIB Working Group 3 'Code of Practice for the selection of main contractors' gives guidance.

2 The number of contractors invited to tender should also relate to the magnitude of the project. Tendering is a costly business which is directly proportional to the size of the project and, as a general rule, the larger the project the fewer should be the number of tenders invited. Full and fair competition must obviously be safeguarded in the interests of the Employer and the construction industry and, where the Employer's programme of works involves a large number of contracts, the operation of a system of selection should ensure that equal opportunities are offered to all qualified contractors commensurate with their size and performance.

Preliminary Enquiries **2.5**

1 With invitations restricted to a small list of contractors drawn from an approved list, it is customary to make preliminary enquiries as to a contractor's desire to tender for each particular contract. The object of this preliminary enquiry is to ensure that the required number of completed tenders is returned and, in the event of a contractor declining, that another can be substituted from the list to make up the desired number. Since the contractor's acceptance of the invitation to tender depends upon the its commitments at the time, declining an invitation is a mark of responsibility, and it should be made clear to contractors so approached that their refusal to tender on the occasion in question will not prejudice their opportunities of tendering on future occasions. Contractors should be informed about the anticipated tender invitation time scales in order that they can plan to have resources available to deal with the tender. The contractors so notified should be advised immediately of any revision of the time scales because they may have to consider, on the strength of the information given, their responses to other opportunities.

2 The description of the Works given in preliminary enquiries and/or advertisements should be in sufficient detail to enable potential tenderers to gauge the scope and the order of cost. Above the threshold level the Works should be advertised under the UK Contracts Regulations or the EC Public Procurement Directives procedures.

3 The information provided should include:

(a) the criteria to be applied in assessing tenders,

(b) the location of the site,

(c) the nature of the work, including whether any Permanent Works design is to be carried out by the Contractor,

(d) the standard Conditions of Contract used together with any significant Special Conditions,

(e) any requirements for bonds or guarantees,

(f) the anticipated contract commencement date,

(g) where known, the required completion date,

(h) any particular sub-contract or supplier requirements of the Employer,

(i) the major quantities involved, and

(j) any special features, e.g. IT requirements, environmental factors.

Period for Tender **2.6**

1 The preparation of tenders is a task of intense activity by contractors and it is in the interests of the Employer to allow adequate time for this from the outset. The tender period should be decided having regard to the value and complexity of the project and the extent to which the tenderers will be involved in negotiations with others. It should not be less than four weeks and for large projects a period of eight to twelve weeks is recommended. Where any substantial or complicated part of the Permanent Works is to be designed by the Contractor, adequate time for design must be included in the tender period. If the actual quantities for such Works are to be paid for under the Contract then the time allowed need only be sufficient for outline design, but if such works are to be covered by a firm or fixed price in the tender then the full design time must be allowed for in the tender period.

2 If, during the tender period, a tenderer submits that insufficient time has been allowed and an extension is granted, all the tenderers must be notified accordingly. In fairness to all tenderers, the 'Instructions to Tenderers' should include a statement to the effect that requests for extension of time will not be considered after a specified date which allows the extension to be notified to tenderers within the first half of the tender period.

Commencement Date 2.7

1 Tenderers should be told at preliminary enquiry and again at tender invitation stages of approximate dates envisaged for tender acceptance and for the Works to be commenced. These dates can be given as an approximate guide only and this should be stated in the 'Instructions to Tenderers'.

3 TENDER DOCUMENTS

Instructions to Tenderers **3.1**

1 The 'Instructions to Tenderers' are issued with the tender documents to draw attention to the conditions which apply to the invitation and the procedures to be followed or complied with by the contractor when compiling and submitting his tender. They should be unambiguously drafted so that nothing within them could affect the rates or prices inserted in the tender or bear upon any other contractual matter and are given solely for the purpose of conveying information and instructions which apply during the tender or pre-contract period. This section is also the framework for incorporating the regulations and procedures imposed by the corporate structure of many Employers and funding bodies, together with the principles and objective criteria used in assessing the most favourable tender. The Instructions should stress the importance of all entries and signatures being unambiguous and that any necessary alterations must be initialled. It should be made clear at the time of tender, by reference in a letter of Invitation to Tender, whether the 'Instructions to Tenderers' will form part of the contract.

2 The 'Instructions to Tenderers' should include **a list of all items required to be included with the tender** and should, in addition, draw attention to such matters as:

(a) The inclusion or otherwise of a Contract Price Fluctuations Clause.

(b) What parts of the Works, if any, are to be designed by the Contractor.

(c) Any special Quality Assurance procedures required.

(d) Any particular submission relating to environmental issues, energy considerations and working methods, e.g. environmental impact assessment, proposed energy audit.

(e) Requirements in respect of safety (including availability of a Health and Safety Plan and Files).

(f) The name of the Planning Supervisor under the CDM Regulations.

(g) The name of the Principal Contractor under the CDM Regulations (if known) or whether the successful tenderer is required so to act.

(h) Bond and/or Parent Company guarantee requirements.

(i) Insurance requirements/procedures.

(j) The means of dealing with queries, any pre-tender meetings and arrangements for inspecting the site.

(k) The acceptability or otherwise of qualified tenders.

(l) The acceptability or otherwise of alternative proposals.

(m) How errors will be dealt with.

(n) Any matters that may condition the Employer's decision to enter into a contract, and the criteria which will be used in assessing offers during the tender adjudication.

(o) The procedures for publishing the tender results.

(p) Requirements for presenting rates and prices in the Bills of Quantities e.g. the use of decimal places.

(q) Requirements for Project Management of the contract and any particular management arrangements.

(r) Acknowledgement of receipt of documents.

(s) Procedures to be adopted in presenting and submitting tenders.

(t) Venue, time and date for return of tenders and applications for an extension of time for tendering and any period for acceptance of the tender which the Employer has determined.

(u) Approximate commencement date envisaged and completion times.

(v) Arrangements for inspecting the Site and any particular documentation during the tender period.

(w) Any special tax considerations.

(x) Reference to any other documentation or certification which may be required to validate the offer.

(y) The period for which the tender is to remain open for acceptance.

(z) That the Employer will not be bound to accept the lowest or any tender.

Conditions of Contract **3.2**

1 Employers are recommended to use one of the current standard Conditions of Contract appropriate to the type of work to be carried out without alteration. Reference to amendments or other matters should be to a specific issue of the Conditions of Contract and not just the 'latest edition'.

2 (a) Each set of Conditions has been prepared after careful drafting and with legal advice; many of the clauses are inter-dependent. It is emphatically stressed that alterations to the text should be avoided wherever possible because this may cause ambiguities and/or errors to arise and may also destroy the balance of the contract as between Contractor and Employer.

(b) It has been the policy of the CCSJC to interpret requirements imposed by Government legislation in standard contract documents only where this has a direct bearing on the relationship between the Employer and the Contractor. The standard Conditions of Contract require the Contractor to comply with Statute but do not tell him how to achieve compliance under the Contract. However, there may be instances where the Contract sets a specific application of Statute, for example noise or pollution levels. Such criteria would be included in the 'General Requirements' section of the Specification.

(c) Conditions of Contract, Bills of Quantities or Payment Schedules and Specifications should be kept separate and amendments to standard Conditions of Contract should never be included in the Specification and vice versa.

(d) Where Special Conditions of Contract for special needs are considered necessary these should be incorporated in accordance with the Guidance Notes to the relevant edition of the ICE Conditions of Contract.

3 When issuing Tender Documents the Employer / Engineer must take particular care to see that the Appendix to the Form of Tender (Part 1) (or for the Minor Works the Appendix to the Conditions of Contract) is fully and carefully completed, together with appropriate entries in any 'Contract Price Fluctuation' clauses to be included.

Site Information 3.3

All factual information related to the Site and the ground conditions established by the Engineer and/or the Employer for the purpose of his design should be included in the tender documents or otherwise made available for inspection by the tenderers. This may include such items as site history, borehole logs, contamination, rainfall records, river levels, tidal records, earthquake information and any special requirements or restrictions that may need to be considered. If demolition is involved then the property safety file must be made available. The information should be given without comment, allowing the tenderer to draw his own conclusions insofar as they may affect his tender. Where geotechnical reports have been made which include test results or other derived material, these should, if relevant, be included with the tender documents or where this is not possible be made available for inspection, although it should be made clear that the Employer accepts responsibility only for the accuracy of the factual information supplied.

Supply of Documents 3.4

1 The number of sets of tender documents supplied to the tenderer should have regard to the complexity of the Works and the involvement of specialist sub-contractors. This should normally be not less than two sets, one copy of the priced Bills of Quantities (or the pricing schedule with the Contract Price if it is a Lump Sum or Design and Construct Contract) being returned with the tender. Sufficient copies should also be supplied to the Employer at the outset, to ensure that ample copies are available for immediate reference when required. Hard copy of tender documents should always be issued for invitation and offer. The issue of documents in electronic form requires particular consideration with regard to licences and passwords. However, the use of electronic data is considered to be essential for documents of the length and complexity of many civil engineering tenders and contracts. In particular, drawings and Bills of Quantities (or pricing schedules) issued electronically can be of great aid to tenderers but it should remain a requirement that the offer documents are hard copy.

2 Drawings (including bar bending schedules where available) should always be issued with the tender documents and should not be made available for inspection only; the estimators need these drawings in order to price the Works and the drawings usually contain important notes of requirements. Notes on drawings are there to amplify the design details; they should always be clearly set out in print size that is easily read, and should not conflict with or be used to vary the Specification. Where not available in electronic form it is also good practice to provide the tenderer with a set of reduced sized drawings in addition to the full sized set and for copy negatives to be made available should the contractor require them.

3 Drawings and data issued for information only and to enable the tenderer to understand more clearly the project requirements should be marked 'For Information purposes - Non contractual document'. Examples of such information could be construction on adjacent sites, plant to be positioned as a subsequent operation, programme of works for public utilities.

4 Where selective tendering procedures have been followed it is inappropriate to require deposits for tender documents. For open tendering a deposit sufficient to discourage idle enquiries may be appropriate. However tender documents may be numbered and tenderers may be asked to acknowledge receipt of them.

4 TENDER PERIOD

Queries 4.1

1 A tenderer should endeavour to clear up all points of doubt before submitting his tender, so that he may present an unqualified tender. Queries should be raised as early as possible and it is desirable to impose a time limit within the tender period for their receipt.

2 Queries of a very minor factual nature which can be dealt with by the Engineer referring the tenderer to some particular Specification clause or detail on a drawing, need not be repeated to all tenderers but the answer given should be confirmed in writing as a non committing statement direct to the tenderer concerned. Where queries concern possible ambiguities, the need for supplementary information or errors in the documents, any verbal replies must be recorded and then confirmed in writing by way of addenda notices or amendments issued to all the tenderers. It is important to clear any points of doubt as soon as they arise.

Site visits by Tenderers 4.2

1 It is generally a requirement of the 'Instructions to Tenderers' that the Contractor shall visit the site and arrangements for this are set out. Tenderers should be advised of the probable duration of the visit, the periods when visits are possible and the maximum number of people who should attend. The visits should take place about half way through the tender period and should be for inspection purposes. They should not be considered as a pre tender meeting but a record of the visit should be kept by the Engineer.

2 Tenderers would not normally be expected to undertake physical tests or to uncover surfaces during the tender period. Special arrangements for obtaining additional information (which may have particular significance to the tenderer in a Design and Construct contract), need to be treated with care to ensure that fairness and confidentiality are preserved in the tender period.

Pre-tender Meetings 4.3

1 Should the Employer and/or his Project Manager be represented at a pre-tender meeting his role with respect to. that of the Engineer needs to be clearly understood by all involved. Where pre-tender meetings are necessary they should be arranged in such a way that they do not prejudice the fairness of competitive tendering and should be offered to all tenderers. Such meetings may be helpful in clarifying points of doubt and uncertainty that may have arisen during the tender period. They may be held collectively with all tenderers present or on an individual basis.

2 Pre-tender meetings should be used as a means to clarify any issues relating to the Tender, and to discuss any appropriate alternatives, sub-contracting of parts of the Works or input from subcontractors. For individual meetings, the use of a standard agenda is essential to observe impartiality. All information provided on such occasions must be properly recorded and confirmed in writing to all tenderers and treated as confidential. Where this information has contractual significance it must be incorporated into the contract documentation itself.

Amendments 4.4

1 Any amendment to the tender documents made by the Engineer/Employer after their issue is disruptive to the tendering processes and imposes a substantial additional burden upon the tenderers. Amendments should be issued as early as possible and only if the Engineer is convinced that the priced tenders will be significantly different from the actual work now to be undertaken. Unless significant changes have to be made, amendments are best withheld and dealt with either as post tender clarification or as part of the administration of the contract after acceptance.

2 Tenderers must be notified immediately if it is decided to issue any amendments to the documents. Where the amendments are other than minor ones an extension to the tender period, appropriate to the situation thereby created, should be considered.

3 Amendments to the documents should be sent out by means of addenda letters or notices which must be numbered consecutively and accompanied by an acknowledgement slip. The use of corrigenda sheets may also be found useful in coping with such situations. Tenderers should be instructed to return counter-signed copies of each letter or notice with their tender, so as to confirm that the addenda were received and acted upon as part of the tender and the contract documents.

Arrangements to Receive Tenders **4.5**

1 Instructions as to the time, date, place, packaging and address format for returning and opening of tenders should be included in the 'Instructions to Tenderers '. Tenders should be sent by registered post, or recorded delivery service, or delivered by hand in a plain sealed envelope, entitled as directed but not bearing any indication of the name of the tenderer. When delivered by hand a dated and timed receipt must be given to the bearer. No discussion should be entered into with the bearer as to the origin or contents of the tender. Envelopes used for the delivery of tender documents must be clearly marked 'Tender document not to be opened until (date and time). '

2 Tenders are confidential and must not have a wider circulation than is necessary for the purpose of deciding the award of the contract. Tenders must therefore be properly safe-guarded at all times and suitable arrangements made to receive and store the documents. These may include briefing ancillary staff on the procedures to be adopted when a tender is delivered, time and date stamping, and the routine to be followed in delivering the unopened tender to the formal recipient who should sign for it. Tenders should remain unopened and be made secure until the appointed time for opening.

3 Additional copies of tenders can be a source of error and should be avoided. Where these are absolutely essential, photocopies or an electronic version of the original tender should be used and should be marked `COPY' on the understanding that the master original tender will take precedence in the event of any discrepancy.

4 Tenderers have the right to modify their tenders in writing at any stage before the appointed submission time. In the event of a tender being delivered before the due submission time, and if the tenderer then wishes to modify it, the tenderer may submit a separate supplementary statement of amendments to his first tender submission. This should also be delivered in a plain sealed envelope for consideration as a supplement at the time of opening tenders and clearly marked and kept in the same way as the original tender.

Communications **4.6**

1 In all communications regarding the enquiry and tender, the use of facsimile facilities is acceptable provided that they are addressed and headed in the same manner as the tender enquiry. Similarly computer print-outs for Bills of Quantities should be accepted. Care should be taken with the confidentiality of information sent by these means.

2 The use of electronic transmission only for tender information which will form the basis of a contract is not acceptable. As a means of making multiple copies of documents discs and CD-ROM will undoubtedly be used. The original document should be a hard copy which may be accompanied by a CD ROM to assist proliferation.

Late Tenders **4.7** Tenders received after the due time and date are invalid and should not be considered. If a late tender is received the tenderer (who may be identified by a process of elimination) should be immediately informed in writing that it has not been admitted because it was received late.

Alternative Tenders **4.8** **1** The 'Instructions to Tenderers ' should state whether offers based on alternative designs will be considered and, if so, should state the minimum design criteria and requirements that will apply as well as the extent to which the requirements specified in the tender documents cannot be changed. The Employer's interests are often best served by facilitating the production of more economic solutions and tenderers should not be unduly constrained.

2 Where alternative offers are permitted and the UK Contracts Regulations or the EC Public Procurement Directives apply, the Employer may not be able to insist that an alternative is accompanied by a tender strictly in accordance with the tender documents. Nevertheless, the Employer is not bound to accept an alternative and for that reason, as well as to preserve parity of tendering, tenderers should be advised to submit a clean compliant tender together with any alternatives they wish to put forward.

3 Where alternative designs are permitted procedural rules should be observed which ensure a fair and proper adjudication of the tenders. These rules should be set out in the 'Instructions to Tenderers ' and might be typically:

(a) If a tenderer wishes to submit an alternative offer this should be made in addition to a tender based strictly on the requirements specified in the tender documents.

(b) A tenderer is to give at least two weeks (or longer) notice before the date of return of tenders of his intention to submit an alternative.

(c) An alternative offer is to be accompanied by supporting information such as drawings, technical assumptions, calculations and a priced Bill of Quantities addendum, in order that its engineering acceptability, construction time and economics can be fully assessed. Documents submitted as alternative designs must be subjected to technical checking and also Quality Assurance procedures which are clearly identified in the offer. Compliance with CDM Regulations, Environmental approvals, time constraints, Planning Permission under the Town and Country Planning Acts and Building Control Regulations approval are issues which can affect the viability of alternatives and the Contractor should obtain any necessary approvals at his own expense.

4 To emphasise the principle of tendering against a conforming design, the 'Instructions to Tenderers' usually contains a statement such as 'The Employer does not bind himself to accept any alternative offer submitted by the tenderer'.

Qualifications **4.9** **1** Unless tenders are submitted strictly in accordance with the tender documents it is very difficult to maintain parity of tendering. Qualifications to a tender undermine this principle and can inhibit fair adjudication by rendering the offer non-comparable with others. They most often arise as proposed changes to the Conditions of Contract or as interpretations of the Specification. The objective should therefore be to reduce the likelihood of qualifications by careful pre-planning and preparation up to and during the tender period, by avoiding ambiguities and, above all, by anticipating the problems of interpretation which tenderers and their suppliers have to resolve when pricing the work.

2 Whilst, as a generality, qualifications should be avoided because they tend to undermine the principle that, whilst parity of tendering must be maintained, there may, on occasions, be overriding circumstances that cause the Engineer to deem it right and in the best interests of the Employer that he should consider and assess a qualification.

3 In order to avoid misunderstandings about the effect that qualifications will have upon the adjudication the 'Instructions to Tenderers ' should make it absolutely clear the circumstances in which qualifications will lead to the tender being rejected, or otherwise how they will be dealt with. It is essential that any such qualifications are properly dealt with prior to acceptance of a tender.

5 TENDER ADJUDICATION

Withdrawal of Tender 5.1

1 Under English Law a tenderer may withdraw his tender at any time up to its formal of acceptance, whereas under Scots Law the tender is deemed to stand firm and is open to acceptance for the period stated in the terms of the offer. The period during which tenders should remain open for acceptance may vary according to the circumstances including the proposed commencement date of the contract and any legal implications of 'fairness' to tenderers. The following general recommendations apply:-

(a) The period for acceptance should be stated.

(b) The period should not be unduly long but must enable proper validation to take place.

(c) Normally a 48 day period is acceptable but it may be increased to between 60 and 120 days to account for the complexity of the tender.

(d) If the tenderer withdraws before the acceptance date then he may not re-submit a tender for that contract.

2 It is essential that everything possible be done to avoid delay in the award of contracts, especially in the case of fixed price contracts. Where the award of the contract is unduly delayed on account of unforeseen and unavoidable circumstances, the most favourably placed tenderers should be advised of any further expected delay and asked whether they will agree to extend the original validity period of their tenders.

Errors 5.2

1 It is essential that the procedures that will be adopted to deal with errors in tenders noted during their examination are described in the 'Instructions to Tenderers'. These procedures should deal equitably with the circumstances arising from errors within the principle that parity of tendering must be maintained.

2 All tenders should be checked for arithmetical accuracy immediately after being opened. The principle to be followed in checking the tender is that rates entered in the Bill of Quantities are the tendered rates and the extension is the product of the rate and item quantity. If the extended sum entered in the Bill of Quantities for each item does not agree with that product it is in error and should be corrected. The summation of the extensions corrected in this way constitutes the total of the priced Bill of Quantities. It is usual to include in the Bill of Quantities an adjustment or balancing item so that last minute changes to the tender sum may be made by tenderers. The method of dealing with this for interim certificates also needs to be made clear. Arithmetic errors in the extension of individual items can be adjusted in the adjustment item so that the tendered total price remains the offer. Alternatively, the 'Instructions to Tenderers' may acknowledge that errors in arithmetic should be corrected and reflected in the offer price used for adjudication. For lump sum contracts, the Lump Sum must stand as the offer.

3 In measure and value contracts such arithmetical errors are automatically eliminated by the remeasurement process, but it is normal courtesy to inform the tenderer of the fact and of the corrected amount of the total of the priced Bill of Quantities immediately the error(s) have been discovered. The foregoing accords with the premise embodied within the ICE Conditions of Contract that the tenderer, in appending rates to the Bill of Quantities, is deemed to have satisfied himself before submitting his tender as to the correctness and sufficiency of the tender rates to cover all obligations under the Contract. Their computation, therefore, in total relative to the billed quantities is the basis for deriving the Contract Price.

4 If a tenderer discovers an error in his tendered rates before the tender date he can amend this by submitting a corrected rate on or before the due date (see 4.5.4). He cannot correct a tender rate after this date when he then has the option of allowing it to stand or withdrawing his tender (except possibly in Scotland). To allow any other approach might allow parity of tendering to be breached.

5 The methods adopted by tenderers to cost and price contracts are many and varied and it is very difficult for the Engineer to establish whether an error has or has not been made. Furthermore the results of tendering cannot be held confidential for very long and parity of tendering will be destroyed if negotiations to confirm or change a rate are undertaken post tender. To negotiate in these circumstances would create de facto two stage tendering and promote the practice of tendering for the negotiating table. This would operate to the disadvantage of more careful tenderers and, in the long run, to all Employers. Accordingly it is not regarded as good practice to negotiate for a change of tendered rate on the grounds that it is in error or inappropriate.

6 If the Engineer considers that a genuine mistake has been made he should seek a confidential meeting with the tenderer who is then asked to amend his tender to eliminate the error in whatever way he chooses. In these circumstances the Engineer may proceed to negotiate a correction only if he is satisfied that parity of tendering would not be breached, having regard to the fact that other errors, of which he is unaware, could be present in every tender. It requires careful judgment by the Engineer to decide whether he should seek to change a tendered rate in one tender because of an error he deems to be genuine, whilst there is a probability that equally genuine errors remain undetected in other tenders; as has been said above, to do so tends to undermine the whole basis of competitive tendering.

Notification of Short-lists **5.3** Tenderers should be notified where they stand in relation to the possibility of their being awarded the contract as soon as the tenders have been checked. This should be done by informing tenderers whose tenders are short-listed for closer examination that this is so and similarly those who are not. Such notification should be within fourteen days from the time of submission of tender; failure to do this is unfair to tenderers who could be holding resources available. In the case of those short-listed it should be made clear that this notification does not constitute a letter of intent. Tenderers advised that they have been short-listed should subsequently be notified immediately a decision to accept a tender has been taken (see 6.1).

Post Tender pre-Contract meetings **5.4**

1 Following the submission and study of the tenders, it may be appropriate for the Employer and Engineer to meet separately the lowest or perhaps even the two or three lowest tenderers. Such meetings should serve only to clarify any points discovered in the scrutiny of the tenders for the purposes of contract administration and adjudication. Subjects for discussion might include:

(a) Clarification

(b) Availability of resources

(c) Proposed site and head office organisation

(d) Construction methods envisaged

(e) Third parties rights, insurances, guarantees, bonds

(f) Qualifications (if any considered worthy of consideration).

(g) Outline and/or detailed designs required to be put forward by the Contractor for Permanent Works or any proposed alternatives.

(h) Supplementary documentation, QA plan, Environmental Policy Statement, Method Statements, Test and Inspection Plan, Resource Scheduling.

2 Detailed minutes of such meetings should be approved by both parties and agreement reached on any matters affecting the eventual contract should be recorded in writing for incorporation into the contract documents. Under no circumstances should it be implied at this stage that it is proposed to accept a particular tender.

Negotiated Change of Tender **5.5**

1 Where there is ample mechanism in the contract for dealing with variations from the tender circumstances and information (apart from the correction of errors in the tender documents (see 5.2 above)), only in exceptional circumstances should negotiations be carried out with a tenderer which would alter his Tender Total prior to tender acceptance. However, if it is considered important to amend a tender before acceptance, then negotiations should be carried out in the first instance with the most favourable tenderer. Only in the event of the Employer considering the price adjustment or other requirement of that tenderer to be unreasonable should negotiations proceed with the next most favourably placed tenderer. It is essential to keep detailed minutes of any discussions held and to obtain verification from the tenderer that he agrees with the record, since these agreements will become contractually binding. Where, however, it is necessary to amend the scope of the Works so substantially that the tenders are no longer a valid basis upon which to make the award, all tenderers should be given the opportunity to submit a new tender for the amended scope of the work. It is stressed that such a procedure is totally exceptional and should only be adopted when no other course is possible.

2 In conducting such negotiations, the guiding principle must be to ensure that the confidentiality and fairness of tendering is preserved, and that no one tenderer is given an unfair advantage over others.

Re-tendering **5.6**

Once tenders have been returned to the Employer, it is not in accordance with good practice and fair dealing to invite re-tenders for substantially the same work. If, exceptionally and for the most compelling reasons, re-tendering is instigated, tenderers should receive a full explanation of the circumstances and reasons which led to this decision. The position with respect to the UK Contracts Regulations and the EC Public Procurement Directives needs special consideration as re-tendering is likely to involve the implementation of the whole procurement procedure again.

Assessment Report **5.7** The Engineer should complete his adjudication by submitting an assessment report to the Employer setting out all the actions taken to scrutinise, check and correct the tenders in accordance with the criteria to be used as set out in the Instructions to Tenderers and giving his conclusions and recommendations to the Employer regarding which tender to accept. A primary objective of the tendering procedure is to identify and award the contract to the most cost effective bidder after allowance has been made for the interest or effect of early payments required. However the Employer's interests are not necessarily best served by selecting the lowest tender. Factors such as cash flow, risk of change in final price, running costs, cost effectiveness, whole life costing may be introduced in assessing the most economically advantageous price for relatively complex contracts. However, for many civil engineering contracts with a select list of contractors who have priced the Engineer's design, the lowest price criteria, after adjustment for dealing with errors as set out in the 'Instructions to Tenderers' will often prevail. However, particular care must be taken when dealing with any criteria which may have been included in the 'Instructions to Tenderers' relating to the basis for an award (see paragraph 3.1.2(p)). In forming his recommendation the Engineer should take into account the various factors established during the adjudication period, such as among other things: the type of plant to be used, the constructional methods proposed and the likely financial out-turn of the contract in relation to the rates quoted. In the case of open tendering the Engineer should also assess the potential ability of the tenderer to complete on time and his financial stability. Where the Engineer is an employee of the Employer, the Employer's representatives responsible for considering tenders should act in the same manner as described in this section for the Engineer.

6 ACCEPTANCE PROCEDURE

Notification of Final Results **6.1** **1** The successful tenderer should be notified immediately it is decided to accept his offer by means of a letter of intent which is normally sent before signing of the agreement. Should the Employer decide not to proceed with the project, for whatever reason, all the tenderers should be informed at the earliest possible time and reasons for this decision given.

2 Once the contract has been let, the unsuccessful tenderers should be individually notified of the decision arising from the tender process, including the amount of the successful bid. All tenderers including the successful tenderer should be supplied with a list of the names of tenderers in alphabetical order, and a separate list of tender totals submitted in ascending order of magnitude, thereby affording individual tenderers the opportunity to observe their ranking without specific reference to the others. Press notices should not be issued until unsuccessful tenderers have been notified.

3 At the time of writing to the unsuccessful tenderers it is customary, for reasons of copyright, to request the return of all drawings and unused documents which were not submitted with their tenders. For major projects, where tenders have required considerable input, the opportunity of a debrief should be offered to the unsuccessful tenderers.

Letter of Intent **6.2** **1** Where it is not immediately possible to issue a formal letter of acceptance to the successful tenderer, the issue of a letter of intent to enter into a contract may in some circumstances serve a useful purpose. When the letter is being issued in order to authorise the Contractor to make preparations for carrying out certain works before formal acceptance of the tender, it can constitute a binding contract in itself where the Employer becomes liable for the financial consequences that arise from it. The letter must therefore be precise as to its intentions and treated as a contractual document (although this may not have the same connotations under the UK Contracts Regulations or the EC Public Procurement Directives).

2 A letter of intent is intended as a prelude to eventually signing the Memorandum of Agreement or formal contract and should contain the following:

(a) A statement that it is intended to accept the tender at some date in the future.

(b) Instructions to proceed (or not to proceed) with the required tasks such as the designing, ordering of materials, letting sub-contracts, etc. and to keep the Engineer informed of the progress in such actions.

(c) A statement that if the tender is subsequently not accepted, or the letter of intent withdrawn, the costs legitimately incurred by the Contractor will be paid by the Employer as if a contract had been entered into.

(d) A limit to financial liability before formal award of the contract and any rates of expenditure agreed to.

(e) A statement that when the contract is formally awarded the provisions of the letter of intent will be void.

(f) A request to the Contractor to acknowledge receipt of the letter of intent and to confirm his acceptance of its conditions.

Letter of Acceptance **6.3**

1 A legally binding contract is established when the tender is formally accepted by or on behalf of the Employer. The terms of this acceptance need to be carefully worded to reflect the tender documents and subsequent meetings and correspondence to the extent that they are intended to become part of the Contract. All tender and pre-contracted documents to form part of the contract should be listed. The signing of a formal Memorandum of Agreement may also be required. However, where a Memorandum of Agreement signed by the Parties does not exist, the Contractor should be required to formally acknowledge his acceptance.

2 It is normal practice for the Engineer to prepare a draft letter of acceptance of the tender, setting out any agreed conditions or provisos which have transpired during the tender period and to send this to the Employer with his assessment report. The letter should include the statement that all subsequent directions concerning the Contract will be issued by 'The Engineer ' under the Contract and that all further correspondence in connection with the Contract should be addressed to him other than any formal requirement in the Contract, but in all cases copies of correspondence are to be sent to the Engineer. The letter of acceptance may alternatively be issued by 'The Engineer', on the written authority of the Employer, provided that the existence of such authority is made known. It is necessary, in either case, that the letter should be an unequivocal acceptance of the final offer.

3 If not previously determined, and in any event before the award of the contract, it is advisable for the Engineer to confirm with the Employer the powers which he intends to delegate to the Engineer's Representative. These are to be notified to the Contractor before the start of (and from time to time during) the contract. Before the award of the Contract it is also advisable for the Engineer to confirm with the Employer the duties and authority which he is to assume under the Contract.

7 CHECK-LIST

Introduction 7.1 The following summary is intended as an action check list for easy reference to the full text of the Guidance.

General 7.2

1 Avoid open tendering if possible but if adopted inform tenderers of the number of contractors to whom the tender documents have been issued.

2 Pre-qualify interested contractors after they have submitted details of their relevant experience, organisation and financial standing. Use Constructionline as appropriate. Be aware of legislation with regard to invitations, tender and award of public body contracts.

3 Draw up a short-list of 4 to 8 tenderers suitable for the contract. Keep standing lists for different values and types of work, and review them annually. To ensure receipt of sufficient tenderers from such lists, write to contractors describing the Works and asking for confirmation that they will tender for a particular contract--making it clear that their refusal will not prejudice their future opportunities.

Tender Documents 7.3

1 Accompany the tender documents with Instructions to Tenderers which should include items in paragraph 3.1.2 (a) to (z) which are relevant.

2 Include in the tender documents, or make available, all known factual information specific to the Site and the ground conditions.

3 Issue at least two copies of the tender documents, including sets of drawings, and more if the work involves specialist sub-contractors; (alternatively, issue or make available information by electronic means and/or copy negatives of the drawings).

4 In preparing Tender Documents attention should be paid to full and careful completion of any Appendix to the Form of Tender or Conditions of Contract.

Tender Period 7.4

1 Allow at least 4 weeks for tendering - more if the contract is large or complex. UK Contract Regulations, EC Public Procurement Directives or particular client procedures may influence the tender programme for public sector works.

2 Impose a time limit for receipt of queries from tenderers.

3 Issue to all tenderers a list of the queries received together with the answers.

4 Hold pre-tender meetings and Site visits, record all information given and confirm in writing to all tenderers.

5 Do not issue amendments to the tender documents unless they will seriously affect the pricing; (minor amendments are best withheld and dealt with at a later date).

6 Keep submitted tenders secure until the opening time (which should be witnessed) and then treat them as confidential documents.

7 Do not accept late tenders (though there may be circumstances where they can be accepted, eg if a facsimile stating the tender sum is received on time and there is clear evidence that the tender was despatched in time).

8 Plan ahead to reduce the need for qualified tenders.

Tender Adjudication 7.5

1 The Employer should aim to place contracts preferably within 2 months of receipt of tenders and within the stated acceptance period. He should particularly avoid delays in acceptance of fixed price tenders. UK Contracts Regulations or EC Public Procurement Directives may reduce this period for public sector work.

2 If errors have been made, assess tender totals on the summation of the extension of the tendered rates and make adjustments as contained in the 'Instructions to Tenderers' if this alters the submitted total the tenderer should be informed.

3 Do not allow tender rates to be altered after opening of the tenders; (in England and Wales a tenderer may withdraw if he is unwilling to stand by a wrong rate).

4 Notify tenderers as soon as possible, preferably within two weeks of tender opening, whether they are short-listed or not.

5 Call meetings with any of the lower tenderers, if necessary, to clarify points of doubt or to check with them their availability of resources, proposed organisation and construction methods. (Minutes of such meetings should be approved by both parties and the agreements reached with the successful tenderer incorporated into the Contract if intended by the Parties).

6 If the Employer wishes to effect a significant change to the scope of the Works, new rates should be negotiated with the favoured tenderer.

7 Re-tenders should only be invited in very exceptional circumstances, in which case tenderers should be notified of the reasons.

8 Do not necessarily select the lowest tender if the Employer's interests might be better met by another when account is taken of construction methods, organisation and likely completion date or financial out-turn. Strictly adhere to any award criteria contained in the 'Instructions to Tenderers'.

Acceptance Procedures 7.6

1 Inform all tenderers as soon as possible of the outcome.

2 Arrange for unsuccessful tenderers to return any retained documents.

3 Issue a letter of acceptance setting out any conditions or provisos which have been agreed since issuing the documents and directing the Contractor to carry out all further correspondence with the 'Engineer ' under the Contract.

4 If there is a delay to formal acceptance, a letter of intent may be sent containing instructions to proceed (or not to proceed) with design, ordering materials etc and setting a limit to financial liability before formal acceptance.

8 BIBLIOGRAPHY

Other Relevant Documents

8.1 Construction Industry Board (CIB) Working Group 3: 'Code of Practice for the Selection of Main Contractors'.

Engineering Construction Industry Association (ECIA): 'Code of Fair Contracting Practice'.

EC/Lloyds Register: 'Guidelines on Risk Issues' 1993.

EC/Lloyds Register: 'Guidelines on Environmental Issues' 1994.

FIDIC: 'Tendering Procedure' 2nd Edition 1994.

Department of the Environment, Transport and the Regions (DETR) 'Rethinking Construction'. The Egan Report 1998.

UK and EC Public Procurement Legislation

8.2 A comprehensive list of UK and EC Public Procurement legislation is available from HM Treasury web site:

http://www.hm-treasury.gov.uk/pub/html/docs/cup/sopdoc.html

Some of the Statutory Instruments on Public Procurement are listed below.

SI 1991 No 2680 – The Public Works Contracts Regulations 1991
SI 1993 No 3228 – The Public Service Contracts Regulations 1993
SI 1995 No 201 – The Public Supply Contracts Regulations 1995
SI 1996 No 2911 – The Utilities Contracts Regulations 1996

ICE Conditions of Contract publications

8.3 The following are the ICE Conditions of Contract publications that are available:

ICE Conditions of Contract Measurement Version 7th Edition
Guidance Notes to the ICE Conditions of Contract 7th Edition
ICE Design and Construct Conditions of Contract
Guidance Notes to the ICE Design and Construct Conditions of Contract
ICE Conditions of Contract for Minor Works 3rd Edition
(including Guidance Notes)

ICE Conditions of Contract

First Edition

CONDITIONS OF CONTRACT AND SCHEDULES TO BE USED IN CONJUNCTION WITH A BI-PARTY CONTRACT

Partnering Addendum

Institution of Civil Engineers
Association of Consulting Engineers
Civil Engineering Contractors Association

May 2003

Published by Thomas Telford Publishing, Thomas Telford Limited, 1 Heron Quay, London E14 4JD on behalf of:

The Director General and Secretary
Institution of Civil Engineers
One Great George Street
London SW1P 3AA

The Chief Executive
Association of Consulting Engineers
Alliance House
12 Caxton Street
London SWlH 0QL

The Director
Civil Engineering Contractors Association
Construction House
56-64 Leonard Street
London EC2A 4JX

The first edition was approved by the Institution of Civil Engineers, the Association of Consulting Engineers and the Civil Engineering Contractors Association, May 2003

The Institution of Civil Engineers, the Association of Consulting Engineers and the Civil Engineering Contractors Association have, as sponsoring authorities, approve the first edition of the document commonly known as the ICE Conditions of Contract, Partnering Addendum, an addendum to the bi-party contract for multi-party partnering. A permanent joint committee has prepared and will keep under review the use of the document and will consider any suggestions for amendment, which should be addressed to the Director General and Secretary (CCSJC), Institution of Civil Engineers, One Great George Street, London SWlP 3AA. Revision to the document will be made when such action seems warranted.

9 8 7 6 5 4 3 2 1

ISBN 0 7277 3231 5

A catalogue record for this book is available from the British Library

Printed and bound in Great Britain

CONTENTS

The ICE Conditions of Contract Partnering Addendum can be used with:

ICE Conditions of Contract Measurement Version 7th Edition (November 1999)
ICE Conditions of Contract Design and Construct 2nd Edition (September 2001)
ICE Conditions of Contract Minor Works 3rd Edition (April 2001)
ICE Conditions of Contract Term Version 1st Edition (September 2002)
ICE Conditions of Contract Ground Investigation (October 1983)

ACE Agreement 2002 - Agreement A(1) Civil/Structural
ACE Agreement 2002 - Agreement A(2) Mechanical/Electrical
ACE Agreement 2002 - Agreement B(1) Civil/Structural
ACE Agreement 2002 - Agreement B(2) Mechanical/Electrical
ACE Agreement 2002 - Agreement C(1) Civil/Structural
ACE Agreement 2002 - Agreement C(2) Mechanical/Electrical
ACE Agreement 2002 - Short Form Agreement
ACE Agreement 2002 - Sub-Consultancy Agreement
ACE Agreement 2002 - ACE/APS Agreement
ACE Conditions of Engagement Agreement F

CECA Form of Sub-Contract for Civil Engineering, Building and Mechanical and Electrical Works (July 1999)
CECA Form of Sub-Contract for use with the ICE Conditions of Contract 6th Edition (July 1998)
CECA Form of Sub-Contract for use with the ICE Conditions of Contract Design and Construct (November 1998)

ACKNOWLEDGEMENTS

The First Edition of the ICE Conditions of Contract Partnering Addendum was produced by the Conditions of Contract Standing Joint Committee. The members of the Partnering Addendum Working Group were:

A. Kondys FRICS
D. Tadiello CEng
R. Beales

The members of the Conditions of Contract Standing Joint Committee are:

D. H. Hodgkinson, (Chairman) MSc CEng FICE MIHT
G. F. Hawker (Legal Advisor) TD BSc(Eng) FEng FICE FIEI
I. Elliott BSc CEng MICE MCIWEM
W. R. Woolley CEng MICE
A. Scott
J. G. Rees BSc(Eng) FICE FCIArb
D. Tadiello CEng
A. Kondys FRICS
R. Beales
R. D. Nokes

Secretariat
F. N. Vernon, (Technical Adviser), BSc, CEng, MICE
J. M. Hawkins, BA (Hons)

Introduction

This Partnering Addendum introduces into a Contract criteria normally associated with multi-party partnering arrangements for a project or programme of works.

This Partnering Addendum can be used with the ICE Conditions of Contract and consultancy and sub-contract agreements produced by the ACE and CECA. With suitably drafted incorporation clauses the Partnering Addendum can be utilised with other forms of contract, including supplier agreements. The single Partnering Addendum is an addendum to each Bi-Party contract. The parties who have this Partnering Addendum included in their contracts are all the parties who are intended to make up the project partnering team. The Partnering Addendum does not create a multi-party contract.

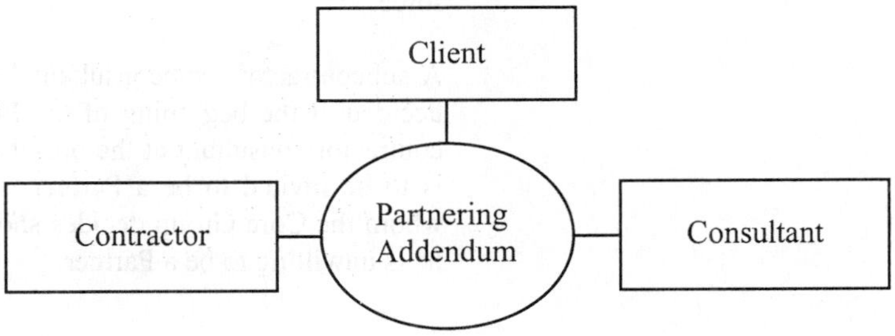

By linking the Partnering Addendum to appropriate Bi-Party contracts, it is intended that the arrangements can be used for:

- partnering for any number of projects (i.e. single project or programme of works)
- the duration of a term contract
- projects nationally and internationally using UK law
- projects of any technical composition
- projects of any size

The partnering arrangement is sufficiently flexible to accommodate:

- additional Partners (disciplines) as the project progresses
- all phases of the project development
- a party entering into more than one Bi-Party contract

The Partnering Addendum is given legal effect by incorporation in each Bi-Party contract and as such comes into effect at the same time as the Bi-Party contract. This Partnering Addendum is not a free standing contract but an addendum that is common to all Bi-Party contracts between Partners. Parties must recognise that by entering into a contract that includes this Partnering Addendum they will be undertaking additional responsibilities.

The Partnering Addendum does not include or create direct remedies between Partners who are not party to the same Bi-Party Contract to recover losses suffered by one Partner caused by a failure of another Partner. Any remedies for such failures are as set out in each Partner's Bi-Party Contract. The existence of the Partnering Addendum and the partnering arrangement will encourage the parties to resolve any differences that arise.

GUIDANCE NOTES

PARTNERS

All significant contributors to the Project should be considered as Partners. As a minimum the Client, the Consultant(s) and the Contractor should be Partners.

The point at which a party becomes a Partner is when his Bi-Party Contract (which includes the Partnering Addendum) comes into existence and he will be named in the Schedule of Partners and his representative identified. Partners shall continue to be bound by the provisions of the Partnering Addendum as long as the Bi-Party Contract, of which it forms part, remains in force.

A subcontractor / subconsultant / supplier may be a Partner. This should be decided at the beginning of the Project. The Core Group should advise the contractor/consultant at the outset if a subcontractor /subconsultant / supplier is to be invited to be a Partner. A subcontractor / subconsultant / supplier whom the Core Group decides should be a Partner should not be appointed if he is unwilling to be a Partner.

CORE GROUP MEMBERS

Not all Partners need to be Core Group members. Partners will be Core Group members for the periods of time that they are key to the project. As a minimum the Core Group should include the Client, the lead Consultant and the main Contractor(s).

The Core Group may invite other parties involved in the project to participate in the Core Group meetings as appropriate.

PARTNERING OBJECTIVES

Experience confirms that Partnering Objectives should be drafted carefully to avoid any conflict with the Bi-Party Contract. The Partnering Objectives can be used to help focus the Partners on the fact that the main driver for setting up the partnering arrangement is to deliver improved performance to the mutual benefit of the Partners. In most cases this will be a simple statement that the objective is to improve on the agreed performance targets embodied in the Partner Incentive Arrangements. If some important issues (e.g. quality, safety) are not a direct part of the Partner Incentive Arrangements the objective(s) will often be written to make it clear that these are not to be sacrificed in pursuit of reward.

Each Partner should work with the other Partners to achieve the Partnering Objectives

PARTNERING OPERATION AND DOCUMENTS

The documents govern the implementation of the partnering arrangement and the relationship between the Partners and may include any documents that are applicable to all parties (e.g. code of conduct) and may include any requirements for:

- use of common information systems, sharing of offices
- attendance at Partners' and Core Group meetings
- participation in partnering workshops
- arrangements for multi-disciplinary design development
- value engineering and value management
- risk management
- joint performance management and monitoring including compliance measurement
- other matters that the Core Group manages

Each Bi-Party Contract should state the deliverables to be provided and required by each Partner and the timing. The Partners will be required to contribute to a consolidated programme showing the detailed timing and sequence of the contribution of all the Partners as set out in the Bi-Party Contracts. Where the agreed consolidated programme changes the requirements in any Bi-Party Contract this shall be dealt with by taking such actions as may be necessary in accordance with their Bi-Party Contract.

The partnering arrangement aims to open up and speed up information flows between the Partners. The Core Group shall agree the transmittal route for information provided.

The Core Group will organise and hold meetings. It will produce and distribute records of each meeting that will include agreed actions. Instructions from the Core Group are issued in accordance with each Partner's Bi-Party Contract.

In order to pool knowledge and facilitate best practice the Partners should give advice and assistance when asked and whenever they identify something that would be helpful to another Partner.

PARTNER RISK MANAGING ARRANGEMENTS

Schedule 5 deals with any allocation of risks that are not in the direct or sole control of one of the Partners that the Partners wish to share.

The identification and allocation of risks is best dealt with in a pre-contract workshop at which the potential Partners identify the risk factors that could impact on each Partner's ability to deliver his portion of the project along with the Partners who are able to control the risk.

When considering any risk managing arrangements the extent of each Partners contribution to the Project should be considered such that any share is commensurate with the input and margin arising from it.

Consultants and contractors with professional indemnity insurance should note that the basis of many professional indemnity policies is negligence and thus will not provide cover for losses payable under risk managing arrangements that are payable in the absence of proven negligence of the consultant or contractor concerned. Policies that are written on a civil liability wording may exclude losses arising from such liability sharing arrangements. Thus the terms of professional indemnity policies should be considered carefully and, if necessary, advice from brokers obtained.

PARTNER KPI/INCENTIVE ARRANGEMENTS

The identification of Key Performance Indicators (KPI) is best dealt with at a pre-contract workshop and is likely to include those required by the Client in relation to his project requirements and those proposed by all potential Partners to promote achievement of the Partnering Objectives and improved performance.

Various incentive arrangements to suit the circumstances of the Project may be used to encourage working together to achieve improvements. These may be linked to certain KPIs and may include:

- Share of under/over-expenditure
- Bonuses for meeting targets

When considering any share of under/over-expenditure the extent of each Partners contribution to the Project should be considered such that any share is commensurate with the input and margin arising from it.

Partners may be best encouraged to work together by a bonus arrangement whereby if a particular target is missed by the poor performance of one or more Partners, then all lose their bonus for that target.

Any KPI/Incentive payment arrangements can be set out in Schedule 6 or each Bi-Party Contract and made known to all Partners.

AMENDMENTS TO SCHEDULES

Where subsequent to the initial partnering arrangement being set up, an additional Partner is to be added it will be necessary to amend the Schedule of Partners and other Schedules as required. The amended Schedules will be incorporated into the Bi-Party Contracts and in turn the partnering arrangement by revision and signature by the Partners of the Schedules in accordance with clause P7.

INCLUDING THE PARTNERING ADDENDUM IN THE BI-PARTY CONTRACTS

By signing the Partnering Addendum the Partners are agreeing to be bound by it. For clarity the Partnering Addendum should be incorporated into the Partner's Bi-Party Contracts as a special clause as follows.

"The Partnering Addendum as amended from time to time by agreement, shall form part of the Bi-Party contract and in the event of ambiguity or discrepancy the Bi-Party contract shall take priority."

The ICE Conditions of Contract Partnering Addendum

Partners P1 The Partners shall be as set out in Schedule 1 as may be amended from time by the Partners in the manner set out in Clause P7.

Core Group Members P2 (1) The Core Group members shall be as set out in Schedule 2 as may be amended from time to time in the manner set out in Clause P7.

(2) The Partners shall select the members of the Core Group. The Core Group shall decide the dates when each member joins and leaves the Core Group. The Client's representative shall be a Core Group member and shall lead the Core Group unless stated otherwise in the Partnering Documents.

(3) The Core Group shall decide how it will operate. Meetings shall be attended by all members or their nominated representatives unless otherwise agreed by all Members. Decisions shall be by unanimous vote of all members in attendance with each member having one vote unless stated otherwise in the Partnering Documents or agreed in writing by all members.

(4) The Core Group shall issue to the Partners copies of Appendix 1 each time it is amended.

Partnering Objectives P3 The objectives of the partnering arrangement are as set out in Schedule 3.

Partnering Operation and Documents P4 (1) The Partnering Documents are referred to in Schedule 4 as may be amended from time to time in the manner set out in Clause P7.

(2) Each Partner shall nominate from time to time a representative to act for him in dealings with other Partners.

(3) The Partners shall work together as a team in accordance with the Partnering Objectives and in the best interests of the project.

(4) The Partners shall use any common information systems as referred to in the Partnering Documents.

(5) Without limitation to the Partner's obligations under the Bi-Party Contracts each Partner shall give a warning to the other Partners as soon as it becomes apparent that

(a) he will be or may become unable to meet his obligations or requirements under his Bi-Party Contract and Partnering Addendum and

(b) he or one of the other Partners will or may be likely to incur delay and or additional costs under his Bi-Party Contract stating the reason, the party or parties responsible and the likely impact

and in each case together with any proposed solutions.

(6) Any difference between Partners shall be notified to the Core Group for consideration. The Core Group shall endeavour to find a mutually acceptable solution.

(7) The Core Group shall prepare and maintain the consolidated programme showing the proposed timing of the contributions of the Partners as set out in the Partners Bi-Party Contracts. The Core Group shall issue a copy of the programme to the Partners and a further copy each time it is revised.

(8) The Core Group shall agree and maintain joint performance monitoring and management procedures as set out in the Partnering Documents and any Partnering KPI/Incentive Arrangements.

(9) A Partner may ask another Partner to provide the information that is needed to carry out the work in his Bi-Party Contract in accordance with the consolidated programme whilst at the same time giving notice of such request to the Core Group. The other Partner shall provide such information to the extent required under his Bi-Party Contract.

(10) Partners shall implement decisions of the Core Group by taking such actions as may be necessary in accordance with their Bi-Party Contract.

(11) Where a Partner issues an instruction under his Bi-Party Contract that may reasonably be expected to have an impact on the other Partners he shall advise the Core Group.

(12) When asked to do so by the Core Group a Partner shall provide such advice information and opinion to the Core Group and to other Partners as may be reasonably required in relation to that that Partner's scope of work under his Bi-Party Contract and the Partnering Objectives.

(13) A Partner shall notify the Core Group of his intention to subcontract any work and the name of the sub-contractor he will use.

(14) Where Partners use a shared project office the provision of and reimbursement for this office and facilities shall become part of the scope of works for one or more of the Partners under the relevant Bi-Party Contract.

Partner Risk Managing Arrangements **P5**

The Partner Risk Managing Arrangements shall as set out in Schedule 5 as may be amended from time to time in the manner set out in clause P7.

Partner KPI/Incentive Arrangements **P6**

(1) Any Partner KPI/Incentive Arrangements shall as set out in Schedule 6.

(2) The implementation of financial adjustments arising from any Partner KPI/Incentive Arrangements shall be under the Bi-Party Contracts.

Schedule Amendments **P7**

The Schedules to this Partnering Addendum may only be amended by the mutual agreement of the Partners.

Status of Partnering Addendum **P8**

(1) This Partnering Addendum shall not (itself) create a legal partnership or any legal relationship between the Partners and does not include direct remedies between Partners who are not party to the same Bi-Party Contract to recover losses suffered by one Partner caused by a failure of another Partner.

(2) This Partnering Addendum does not confer any right on any Partner to represent any other Partner or enter into a commitment on behalf of any other Partner (without such other Partner's prior written consent).

(3) Nothing in this Partnering Addendum confers or purports to confer on any third party any benefit or any right pursuant to the Contracts (Rights of Third Parties) Act 1999 to enforce any term of the Partnering Addendum.

Appendix

Schedules to Partnering Addendum

Schedule 1 - Partners
Schedule 2 - Core Group Members
Schedule 3 - Partnering Objectives
Schedule 4 - Partnering Operation and Documents
Schedule 5 - Partner Risk Managing Arrangements
Schedule 6 - Partner KPI/Incentive Arrangements

Revision No:........................

Latest Revision Date:.............

Approved by Partners:

Partner .... Signature ....

Partner .... Signature ....

Partner .... Signature ....

Partner .... Signature ....

Schedule 1 - Partners

Name of Partner	Address and contact details	Joining date

Schedule 2 - Core Group Members

Name of Partner	Name of member and contact details	Joining date	Leaving date

Schedule 3 - Partnering Objectives

The objectives of the partnering arrangement are:

Schedule 4 - Partnering Operation and Documents

The Partnering Operation and Documents are:

Schedule 5 - Partner Risk Managing Arrangements

The risk managing arrangements between Partners shall be as set out in the Bi-Party Contracts other than as follows:

Schedule 6 - Partner KPI/Incentive Arrangements

The KPI/Incentive arrangements between Partnering Team members are:

ICE Conditions of Contract for Minor Works

Third Edition

CONDITIONS OF CONTRACT, APPENDIX AND FORM
OF AGREEMENT FOR USE IN CONNECTION WITH
MINOR WORKS OF CIVIL ENGINEERING CONSTRUCTION

The Institution of Civil Engineers
Association of Consulting Engineers
Civil Engineering Contractors Association

Published by Thomas Telford Publishing, Thomas Telford Limited, 1 Heron Quay, London E14 4JD on behalf of

The Chief Executive and Secretary
The Institution of Civil Engineers
One Great George Street
London SWIP 3AA

The Chief Executive
Association of Consulting Engineers
Alliance House
12 Caxton Street
London SW1H 0QL

The Director
Civil Engineering Contractors Association
Construction House
56-64 Leonard Street
London EC2A 4JX

The first and second editions were approved by The Institution of Civil Engineers, the Association of Consulting Engineers and the Federation of Civil Engineering Contractors.
First edition 1988
Second edition 1995, reprinted 1998 with an amendment
Third edition was approved by The Institution of Civil Engineers, the Association of Consulting Engineers and the Federation of Civil Engineering Contractors April 2001.

The Institution of Civil Engineers, the Association of Consulting Engineers and the Civil Engineering Contractors Association have, as sponsoring bodies, approved this revised third edition of the document commonly known as the ICE Conditions of Contract for Minor Works, for all minor works of civil engineering construction. A permanent joint committee of the sponsoring bodies has prepared and will keep under review the use of this document and will consider any suggestions for amendment, which should be addressed to the Chief Executive and Secretary (CCSJC), The Institution of Civil Engineers, One Great George Street, London SW1P 3AA. Revision to the document will be made when such action seems warranted.

9 8 7 6 5 4 3 2 1

ISBN 0 7277 2940 3

Printed and bound in Great Britain by Selwood Printing, West Sussex.

CONTENTS

Associated publications referred to in the Contract

ICE Arbitration Procedure 1997
ICE Appendix (2001) to "The Scottish Arbitration Code 1999"
ICE Conciliation Procedure 1999
ICE Adjudication Procedure (1997)

Other ICE Conditions of Contract publications

ICE Conditions of Contract 7th Edition
Guidance Notes to the ICE Conditions of Contract 7th Edition
ICE Design and Construct Conditions of Contract 2nd Edition
Guidance Notes to the ICE Design and Construct Conditions of Contract 2nd Edition

INDEX

ICE Conditions of Contract for Minor Works

1 DEFINITIONS AND INTERPRETATION

Definitions **1.1** "Works" means all the work necessary for the completion of the Contract including any variations ordered by the Engineer.

1.2 "Contract" means the Agreement if any together with these Conditions of Contract the Appendix and other items listed in the Contract Schedule.

1.3 "Cost" (except where otherwise defined for 'cost plus' contracts, see Appendix paragraph 2) means all expenditure properly incurred or to be incurred whether on or off the Site including overhead finance and other charges (including loss of interest) properly allocatable thereto but does not include any allowance for profit.

1.4 "Site" means the lands and other places on under in or through which the Works are to be constructed and any other lands or places provided by the Employer for the purposes of the Contract together with such other places as may be designated in the Contract or subsequently agreed by the Engineer as forming part of the Site.

1.5 "Excepted Risks" are

(a) the use or occupation by the Employer his agents servants or other contractors (not being employed by the Contractor) of any part of the Permanent Works

(b) any fault defect error or omission in the design of the Works (other than a design provided by the Contractor pursuant to his obligations under the Contract)

(c) riot war invasion act of foreign enemies or hostilities (whether war be declared or not)

(d) civil war rebellion revolution insurrection or military or usurped power

(e) ionizing radiations or contamination by radioactivity from any nuclear fuel or from any nuclear waste from the combustion of nuclear fuel radioactive toxic explosive or other hazardous properties of any explosive nuclear assembly or nuclear component hereof

and

(f) pressure waves caused by aircraft or other aerial devices travelling at sonic or supersonic speeds.

Headings and marginal notes **1.6** The headings and marginal notes in the Conditions of Contract shall not be deemed to be part thereof or be taken into consideration in the interpretation or construction thereof or the Contract.

Communications in writing **1.7** Communications which under the Contract are required to be 'in writing' may be hand-written typewritten or printed and sent by hand post telex cable facsimile or other means resulting in a permanent record.

2 ENGINEER

Engineer to be a named individual 2.1 The Employer shall appoint and notify to the Contractor in writing a named individual to act as the Engineer. If at any time the Engineer is unable to continue the duties required by the Contract the Employer shall forthwith appoint a replacement and shall so notify the Contractor in writing.

Delegation 2.2 The Engineer may appoint a named Resident Engineer and/or other suitably experienced person to watch and inspect the Works and the Engineer may delegate to such person in writing any of the powers of the Engineer herein other than those in Clause 11 or Addendum A provided that prior notice in writing is given to the Contractor.

Instructions 2.3 The Engineer shall have power to give instructions for

(a) any variation to the Works including any addition thereto or omission therefrom

(b) carrying out any test or investigation

(c) the suspension of the Works or any part of the Works in accordance with Clause 2.6

(d) any change in the intended sequence of the Works

(e) measures necessary to overcome or deal with any obstruction or condition falling within Clause 3.8

(f) the removal and/or re-execution of any work or materials not in accordance with the Contract

(g) the elucidation or explanation of any matter to enable the Contractor to meet his obligations under the Contract

(h) the exclusion from the Site of any person employed thereon which power shall not be exercised unreasonably.

Exercise of Delegated Powers 2.4 The Engineer or Resident Engineer and/or other suitably experienced person who exercises any delegated power shall upon the written request of the Contractor specify in writing under which of the foregoing powers any instruction is given. If the Contractor is dissatisfied with any such instruction he shall be entitled to refer the matter to the Engineer for his decision pursuant to Clause 11 or Addendum A.

Dayworks 2.5 The Engineer may order in writing that any work shall be executed on a daywork basis. Subject to the production of proper records the Contractor shall then be entitled to be paid in accordance with a Daywork Schedule included in the Contract or otherwise in accordance with the 'Schedule of Dayworks carried out incidental to Contract Work' issued by the Civil Engineering Contractors Association and current at the date the work is carried out.

Engineer may Suspend the Progress of the Works

2.6 (1) The Engineer may order the suspension of the progress of the Works or any part thereof

(a) for the proper execution of the work

(b) for the safety of the Works or any part thereof

(c) by reason of weather conditions

and in such event may issue such instructions as may in his opinion be necessary to protect and secure the Works during the period of suspension.

(2) If permission to resume work is not given by the Engineer within a period of 60 days from the date of the written Order of Suspension then the Contractor may serve a written notice on the Engineer requiring permission to proceed with the Works within 14 days from the receipt of such notice. Subject to the Contractor not being in default under the Contract the Engineer shall grant such permission and if such permission is not granted the Contractor may by a further written notice served on the Engineer elect to treat the suspension where it affects a part of the Works as an omission under Clause 2.3(a) or where the whole of the Works is suspended as an abandonment of the Contract by the Employer.

Parties bound by Engineer's instructions

2.7 Each party shall be bound by and give effect to every instruction or decision of the Engineer unless and until it is altered or amended by agreement or other determination pursuant to Clause 11.

3 GENERAL OBLIGATIONS

Contractor to perform and complete the Works

3.1 The Contractor shall perform and complete the Works and shall (subject to any provision in the Contract) provide all supervision labour materials plant transport and temporary works which may be necessary therefor.

Responsibility for care of the Works

3.2 (1) The Contractor shall take full responsibility for the care of the Works from the starting date until 14 days after the Engineer issues a Certificate of Practical Completion for the whole of the Works pursuant to Clause 4.5.

(2) If the Engineer issues a Certificate of Practical Completion in respect of any part of the Works before completion of the whole of the Works the Contractor shall cease to be responsible for the care of that part of the Works 14 days thereafter and the responsibility of its care shall then pass to the Employer.

(3) The Contractor shall take full responsibility for the care of any outstanding work which he has undertaken to finish during the Defects Correction Period until such outstanding work is complete.

Contractor to repair and make good

3.3 (1) In case any damage loss or injury from any cause whatsoever (save and except the Excepted Risks) shall happen to the Works or any part thereof while the Contractor is responsible for their care the Contractor shall at his own cost repair and make good the same so that at completion the Works shall be in good order and condition and conform in every respect with the requirements of the Contract and the Engineer's instructions.

(2) To the extent that any damage loss or injury arises from any of the Excepted Risks the Contractor shall if required by the Engineer repair and make good the same at the expense of the Employer.

(3) The Contractor shall also be liable for any damage to the Works occasioned by him in the course of any operations carried out by him for the purpose of completing outstanding work or complying with his obligations under Clauses 4.7 and 5.2.

Contractor's authorized representative

3.4 The Contractor shall notify the Engineer of the person duly authorized to receive instructions on behalf of the Contractor.

Setting out and safety of site operations

3.5 The Contractor shall take full responsibility for the setting out of the Works and for the adequacy stability and safety of his site operations and methods of construction.

Engineer to provide necessary information

3.6 Subject to Clause 3.5 the Engineer shall be responsible for the provision of any instructions drawings or other information necessary for the proper and adequate construction and completion of the Works.

Contractor's responsibility for design

3.7 (1) The Contractor shall not be responsible for the design of the Works except where expressly stated in the Contract.

(2) The Contractor shall be responsible for the design of any temporary works other than temporary works designed by the Engineer.

(3) The Contractor shall exercise all reasonable skill care and diligence in designing any part of the permanent Works for which he is responsible.

Adverse physical conditions and artificial obstructions—delay and extra cost

3.8 (1) If during the carrying out of the Works the Contractor encounters any artificial obstruction or physical condition (other than a weather condition or a condition due to weather) which obstruction or condition could not in his opinion reasonably have been foreseen by an experienced contractor the Contractor shall as early as practicable give written notice thereof to the Engineer.

(2) If in the opinion of the Engineer such obstruction or condition could not reasonably have been foreseen by an experienced contractor then the Engineer shall certify a fair and reasonable sum and the Employer shall pay such sum to cover the cost of performing any additional work or using any additional plant or equipment together with a reasonable percentage addition in respect of profit as a result of

(a) complying with any instructions which the Engineer may issue

and/or

(b) taking proper and reasonable measures to overcome or deal with the obstruction or condition in the absence of instructions from the Engineer

together with such sum as shall be agreed as the additional cost to the Contractor of the delay or disruption arising therefrom. Failing agreement of such sums the Engineer shall determine the fair and reasonable sum to be paid.

Facilities for other contractors

3.9 The Contractor shall in accordance with requirements of the Engineer afford reasonable facilities for any other contractor employed by the Employer and for any other properly authorized authority employed on the Site.

4 STARTING AND COMPLETION

Starting date to be notified in writing

4.1 The starting date shall be the date specified in the Appendix or if no date is specified a date to be notified by the Engineer in writing being within a reasonable time and in any event within 28 days after the date of acceptance of the Tender. The Contractor shall begin the Works at or as soon as reasonably possible after the starting date.

Period for Completion

4.2 The period or periods for completion shall be as stated in the Appendix or such extended time as may be granted under Clause 4.4 and shall commence on the starting date.

Contractor's programme

4.3 The Contractor shall within 14 days after the starting date if so required provide a programme of his intended activities. The Contractor shall at all times proceed with the Works with due expedition and reasonably in accordance with his programme or any modification thereof which he may provide or which the Engineer may request.

Extension of period for completion

4.4 If the progress of the Works or any part thereof is delayed for any of the following reasons

(a) an instruction given under Clause 2.3(a)(c) or (d)

(b) an instruction given under Clause 2.3(b) where the test or investigation fails to disclose non-compliance with the Contract

(c) encountering an obstruction or condition falling within Clause 3.8 and/or an instruction given under Clause 2.3(e)

(d) delay in receipt by the Contractor of necessary instructions drawings or other information

(e) failure by the Employer to give adequate access to the Works or possession of land required to perform the Works

(f) delay in receipt by the Contractor of materials to be provided by the Employer under the Contract

(g) exceptional adverse weather

(h) any delay impediment prevention or default by the Employer

(i) other special circumstances of any kind whatsoever outside the control of the Contractor

then provided that the Contractor has taken all reasonable steps to avoid or minimize the delay the Engineer shall upon a written request by the Contractor promptly by notice in writing grant such extension of the period for completion of the whole or part of the Works as may in his opinion be reasonable. The extended period or periods for completion shall be subject to regular review provided that no such review shall result in a decrease in any extension of time already granted by the Engineer.

Certificate of Practical Completion of Works or part of Works

4.5 (1) Practical completion of the whole of the Works shall occur when the Works reach a state when notwithstanding any defect or outstanding items therein they are taken or are fit to be taken into use or possession by the Employer.

(2) Similarly practical completion of part of the Works may also occur but only if it is fit for such part to be taken into use or possession independently of the remainder.

(3) The Engineer shall upon the Contractor's request promptly certify in writing the date on which the Works or any part thereof has reached practical completion or otherwise advise the Contractor in writing of the work necessary to achieve such completion.

Liquidated damages 4.6 If by the end of the period or extended period or periods for completion the Works have not reached practical completion the Contractor shall be liable to the Employer in the sum stated in the Appendix as liquidated damages for every week (or *pro rata* for part of a week) during which the Works so remain uncompleted up to the limit stated in the Appendix. Similarly where part or parts of the Works so remain uncompleted the Contractor shall be liable to the Employer in the sum stated in the Appendix reduced in proportion to the value of those parts which have been certified as complete provided that the said limit shall not be reduced.

Provided that if after liquidated damages have become payable in respect of any part of the Works the Engineer issues a variation order under Clause 2.3(a) or an artificial obstruction or physical condition within the meaning of Clause 3.8 is encountered or any other situation outside the Contractor's control arises any of which in the opinion of the Engineer results in further delay to that part of the Works

(a) the Engineer shall so inform the Contractor and the Employer in writing

and

(b) the Employer's further entitlement to liquidated damages in respect of that part of the Works shall be suspended until the Engineer notifies the Contractor and the Employer in writing that the further delay has come to an end.

Such suspension shall not invalidate any entitlement to liquidated damages which accrued before the period of delay started to run and any monies deducted or paid in accordance with this Clause may be retained by the Employer without incurring interest thereon under Clause 7.8.

Rectification of defects 4.7 The Contractor shall rectify any defects and complete any outstanding items in the Works or any part thereof which reach practical completion promptly thereafter or in such a manner and/or time as may be agreed or otherwise accepted by the Engineer. The Contractor shall maintain any parts which reach practical completion in the condition required by the Contract until practical completion of the whole of the Works fair wear and tear excepted.

5 DEFECTS

Definition of Defects Correction Period 5.1 'Defects Correction Period' means the period stated in the Appendix which period shall run from the date certified as practical completion of the whole of the Works.

Cost of remedying defects 5.2 If any defects appear in the Works during the Defects Correction Period which are due to the use of materials or workmanship not in accordance with the Contract the Engineer shall give written notice thereof and the Contractor shall make good the same at his own cost.

Remedy for Contractor's failure to correct defects 5.3 If any such defects are not corrected within a reasonable time by the Contractor the Employer may after giving 14 days written notice to the Contractor employ others to correct the same and the cost thereof shall be payable by the Contractor to the Employer.

Engineer to certify completion 5.4 Upon the expiry of the Defects Correction Period and when any outstanding work notified to the Contractor under Clause 5.2 has been made good the Engineer shall upon the written request of the Contractor certify the date on which the Contractor completed his obligations under the Contract to the Engineer's satisfaction.

Unfulfilled obligations 5.5 Nothing in Clause 5 shall affect the rights of either party in respect of defects appearing after the Defects Correction Period.

6 ADDITIONAL PAYMENTS

Engineer to Determine additional sums and deductions 6.1 If the Contractor carries out additional works or incurs additional cost including any cost arising from delay or disruption to the progress of the Works as a result of any of the matters referred to in paragraphs (a) (b) (d) (e) (f) or (h) of Clause 4.4 the Engineer shall certify and the Employer shall pay to the Contractor such additional sum as the Engineer after consultation with the Contractor considers fair and reasonable. Likewise the Engineer shall determine a fair and reasonable deduction to be made in respect of any omission of work.

Valuation of additional work 6.2 In determining a fair and reasonable sum under Clause 6.1 for additional work the Engineer shall have regard to the prices contained in the Contract.

7 PAYMENT

Valuation of the Works 7.1 The Works shall be valued as provided for in the Contract.

Monthly statements 7.2 Unless otherwise agreed the Contractor shall submit to the Engineer at intervals of not less than one month commencing within one month after the starting date of the Works a statement showing the estimated value of the Works executed up to the end of that period together with a list of any goods or materials delivered to the Site and their value and any other items which the Contractor considers should be included in an interim certificate.

Interim payments 7.3 Within 25 days of the delivery of such statement the Engineer shall certify and within 28 days of the same date the Employer shall pay to the Contractor (after deducting any previous payments on account) such sum as the Engineer considers is properly due less retention at the rate of and up to the limit set out in the Appendix. Until practical completion of the whole of the Works the Engineer shall not be required to certify any payment less than the sum stated in the Appendix as the minimum amount of interim certificate. The Engineer may by any certificate delete correct or modify any sum previously certified by him.

The payments become due on certification with the final date for payment being 28 days after the date of delivery of the Contractor's monthly statement.

Payment of retention money 7.4 One half of the retention money shall be certified by the Engineer and paid to the Contractor within 14 days after the date on which the Engineer issues a certificate of practical completion of the whole of the Works.

7.5 The final date for payment of the remainder of the retention money to be paid to the Contractor is 14 days after the issue of the Engineer's certificate under Clause 5.4.

Contractor to submit final account 7.6 Within 28 days after the issue of the Engineer's certificate under Clause 5.4 the Contractor shall submit a final account to the Engineer together with any documentation reasonably required to enable the Engineer to ascertain the final contract value. Within 42 days after the receipt of this information the Engineer shall issue the final certificate. The amount payable by the Employer shall become due on certification. The final date for payment is 14 days later.

7.7 The final certificate shall save in the case of fraud or dishonesty relating to or affecting any matter dealt with in the certificate be conclusive evidence as to the sum due to the Contractor under or arising out of the Contract (subject only to Clause 7.11) unless either party has within 28 days after the issue of the final certificate given notice pursuant to Clause 11 or Addendum A.

Interest on overdue payments

7.8 In the event of failure by the Engineer to certify or the Employer to make payment in accordance with the Contract or any decision of an adjudicator or any finding of an arbitrator to such effect the Employer shall pay to the Contractor interest compounded monthly for each day on which any payment is overdue or which should have been certified and paid at a rate equivalent to 2% per annum above the base lending rate of the bank specified in the Appendix.

Certificates and payment notices

7.9 Every certificate issued by the Engineer pursuant to this Clause shall be sent to the Employer and on the Employer's behalf to the Contractor. By this certificate the Employer shall give notice to the Contractor specifying the amount (if any) of the payment proposed to be made and the basis on which it was calculated.

Notice of intention to withhold payment

7.10 Where a payment under Clause 7.3 or 7.6 is to differ from that certified or the Employer is to withhold payment after the final date for payment of a sum due under the Contract the Employer shall notify the Contractor in writing not less than one day before the final date for payment specifying the amount proposed to be withheld and the ground for withholding payment or if there is more than one ground each ground and the amount attributable to it.

Value Added Tax

7.11 (1) The Contractor shall be deemed not to have allowed in his tender for the tax payable by him as a taxable person to the Commissioners of Customs and Excise being tax chargeable on any taxable supplies to the Employer which are to be made under the Contract.

(2) All certificates issued by the Engineer under Clauses 7.3 to 7.7 shall be net of Value Added Tax.

(3) In addition to the payments due under such certificates the Employer shall separately identify and pay to the Contractor any Value Added Tax properly chargeable by the Commissioners of Customs and Excise on the supply to the Employer of any goods and/or services by the Contractor under the Contract.

8 ASSIGNMENT AND SUB-CONTRACTING

Assignment

8.1 (1) Neither the Employer nor the Contractor shall assign the Contract or any part thereof or any benefit or interest therein or thereunder without the written consent of the other party which consent shall not unreasonably be withheld.

Third party rights

(2) Nothing in this contract confers or purports to confer on any third party any benefit or any right pursuant to the Contracts (Rights of Third Parties) Act 1999 to enforce any term of the Contract.

No sub-contracting without Engineer's Consent

8.2 The Contractor shall not sub-contract the whole of the Works. The Contractor shall not sub-contract any part of the Works without the consent of the Engineer which consent shall not unreasonably be withheld.

Contractor responsible for sub-contractors

8.3 The Contractor shall be responsible for any acts defaults or neglects of any sub-contractor his agents servants or workmen in the execution of the Works or any part thereof as if they were the acts defaults or neglects of the Contractor.

9 STATUTORY OBLIGATIONS

Contractor to comply with statutory requirements

9.1 The Contractor shall subject to Clause 9.3 comply with and give all notices required by any statute statutory instrument rule or order or any regulation or by-law applicable to the construction of the Works (hereinafter called 'the statutory requirements') and shall pay all fees and charges which are payable in respect thereof.

Employer to obtain consents

9.2 The Employer shall be responsible for obtaining in due time any consent approval licence or permission but only to the extent that the same may be necessary for the Works in their permanent form.

Contractor's exemption from liability to comply with statutes

9.3 The Contractor shall not be liable for any failure to comply with the statutory requirements where and to the extent that such failure results from the Contractor having carried out the Works in accordance with the Contract or with any instruction of the Engineer.

10 LIABILITIES AND INSURANCE

Insurance of the Works

10.1 (1) If so stated in the Appendix the Contractor shall maintain insurance in the joint names of the Employer and the Contractor in respect of the Works (including for the purpose of Clause 10 any unfixed materials or other things delivered to the Site for incorporation therein) to their full value against all loss or damage from whatever cause arising (other than the Excepted Risks) for which he is responsible under the terms of the Contract.

(2) Such insurance shall be effected in such a manner that the Employer and Contractor are covered for the period stipulated in Clause 3.2 and are also covered for loss or damage arising during the Defects Correction Period from such cause occurring prior to the commencement of the Defects Correction Period and for any loss or damage occasioned by the Contractor in the course of any operation carried out by him for the purpose of complying with his obligations under Clauses 4.7 and 5.2.

(3) The Contractor shall not be liable to insure against the necessity for the repair or reconstruction of any work constructed with materials or workmanship not in accordance with the requirements of the Contract.

(4) Any amounts not insured or not recovered from insurers whether as excesses carried under the policy or otherwise shall be borne by the Contractor or the Employer in accordance with their respective responsibilities under Clauses 3.2 and 3.3.

Contractor to indemnify the Employer

10.2 The Contractor shall indemnify and keep the Employer indemnified against all losses and claims for injury or damage to any person or property whatsoever (save for the matters for which the Contractor is responsible under Clause 3.2) which may arise out of or in consequence of the Works and against all claims demands proceedings damages costs charges and expenses whatsoever in respect thereof or in relation thereto subject to Clauses 10.3 and 10.4.

10.3 The liability of the Contractor to indemnify the Employer under Clause 10.2 shall be reduced proportionately to the extent that the act or neglect of the Engineer or the Employer his servants his agents or other contractors not employed by the Contractor may have contributed to the said loss injury or damage.

10.4 The Contractor shall not be liable for or in respect of or to indemnify the Employer against any compensation or damage for or with respect to

(a) damage to crops being on the Site (save in so far as possession has not been given to the Contractor)

(b) the use or occupation of land (which has been provided by the Employer) by the Works or any part thereof or for the purpose of constructing completing and maintaining the Works (including consequent loss of crops) or interference whether temporary or permanent with any right of way light air or water or other easement or quasi easement which are the unavoidable result of the construction of the Works in accordance with the Contract

(c) the right of the Employer to construct the Works or any part thereof on over under in or through any land

(d) damage which is the unavoidable result of the construction of the Works in accordance with the Contract

(e) death or injury to persons or loss of or damage to property resulting from any act or neglect or breach of statutory duty done or committed by the Engineer or the Employer his agents servants or other contractors (not being employed by the Contractor) or for or in respect of any claims demands proceedings damages costs charges and expenses in respect thereof or in relation thereto.

Employer to indemnify Contractor

10.5 The Employer shall indemnify and keep indemnified the Contractor from and against all claims demands proceedings damages costs charges and expenses in respect of the matters referred to in Clause 10.4. Provided always that the Employer's liability to indemnify the Contractor under paragraph (e) of Clause 10.4 shall be reduced proportionately to the extent that the act or neglect of the Contractor or his sub-contractors servants or agents may have contributed to the said injury or damage.

Employer to approve insurance

10.6 The Contractor shall throughout the execution of the Works maintain insurance against damage loss or injury for which he is liable under Clause 10.2 subject to the exceptions provided by Clauses 10.3 and 10.4. Such insurance shall be effected with an insurer and in terms approved by the Employer (which approval shall not be unreasonably withheld) for at least the amount stated in the Appendix. The terms of such insurance shall include a provision whereby in the event of any claim in respect of which the Contractor would be entitled to receive indemnity under the policy being brought or made against the Employer the insurer will indemnify the Employer against any such claims and any costs charges and expenses in respect thereof.

Contractor to produce policies of insurance

10.7 Both the Employer and the Contractor shall comply with the terms of any policy issued in connection with the Contract and shall whenever required produce to the Employer the policy or policies of insurance and the receipts for the payment of the current premiums.

11 DISPUTES

Settlement of disputes

11.1 If any dispute or difference of any kind whatsoever shall arise between the Employer and the Contractor in connection with or arising out of the Contract or the carrying out of the Works (excluding a dispute under Clause 7.11 but including a dispute as to any act or omission of the Engineer) whether arising during the progress of the Works or after their completion it shall be settled according to the provisions in Addendum A.

12 APPLICATION TO SCOTLAND AND NORTHERN IRELAND

Application to Scotland

12.1 If the Works are situated in Scotland (and unless the Contract otherwise provides) the Contract shall in all respects be construed and operate as a Scottish contract and shall be interpreted in accordance with Scots Law and the provisions of sub-clause (2) of this Clause shall apply.

12.2 In the application of these Conditions and in particular Addendum A hereof

(a) any reference to arbitration under this Clause shall be conducted in accordance with the law of Scotland "The Scottish Arbitration Code 1999" prepared by the Scottish Council for International Arbitration, the Chartered Institute of Arbitrators (Scottish Branch) and the Scottish Building Contract Committee together with the ICE Appendix (2001) thereto or any amendment to or modification of the Appendix being in force at the time of appointment of the arbitrator. Such arbitrator shall have full power to open up review and revise any decision, opinion, instruction, direction, certificate or valuation of the Engineer or an adjudicator

(b) for any reference to the "Notice to Refer" there shall be substituted reference to the 'Notice of Arbitration'

(c) the existing Addendum Clause A.10 shall be deleted and replaced with the new Addendum Clause A.10 below

" Appointment of arbitrator

A.10 (a) The arbitral tribunal shall be appointed by agreement of the parties.

President or Vice-President to act

(b) Failing agreement of the parties as aforesaid at sub-clause (a) above the following shall apply.

(i) Reference to Articles 3.5 and 3.6 of the Code to the Chairman of the Institute of Arbitrators (Scottish Branch) and to the Chairman of the Scottish Council for International Arbitration shall be deemed to be a reference to the President of the Institution of Civil Engineers as defined at (ii) below.

(ii) "President" means the President for the time being of the Institution of Civil Engineers or any Vice President acting on his behalf or such other person as may have been nominated in the arbitration agreement to appoint the arbitrator in default of agreement between the parties.

"

(d) the existing Addendum Clause A.11 shall be deleted and replaced with new Addendum Clause A.11 below

" Arbitration – procedure and powers

A.11 (a) Neither party shall be limited in the arbitration to the evidence or arguments put to the Engineer or to any adjudicator pursuant to Addendum Clause A.2 or A.6 respectively.

(b) Unless the parties otherwise agree in writing any reference to arbitration may proceed notwithstanding that the Works are not then complete or alleged to be complete.

"

(e) notwithstanding any of the other provisions of these Conditions or of the Code (including in particular Articles 22.7 and 22.8) nothing therein shall be construed as excluding recourse to the Court of Sessions under Section 3 of the Administration of Justice of (Scotland) Act 1972

(f) where the Employer or the Contractor wishes to register the award of an arbitrator or the decision of an adjudicator in the Books of Council and Session for preservation and execution the other party shall on being requested to do so forthwith consent to such registration by subscribing the said decision before a witness and

(g) the existing Appendix to the Conditions of Contract Item 19 shall be deleted.

(h) The application of Articles 1.6 1.8 1.9 and 1.10 of the Scottish Arbitration Code 1999 shall be subject to the provisions of Addendum Clause A.2 of the Contract unless the Arbitrator otherwise directs.

Application to Northern Ireland **12.3** If the Works are situated in Northern Ireland the Contract shall in all respects be construed and operate as a Northern Irish contract and shall be interpreted in accordance with the law of Northern Ireland.

13 CONSTRUCTION (DESIGN AND MANAGEMENT) REGULATIONS 1994

Definitions **13.1** In this Clause

(a) 'the Regulations' means the Construction (Design and Management) Regulations 1994 or any statutory re-enactment or amendment thereof for the time being in force

(b) 'Planning Supervisor' and 'Principal Contractor' mean the persons so described in regulation 2(1) of the Regulations

(c) 'Health and Safety Plan' means the plan prepared by virtue of regulation 15 of the Regulations

Action to be taken **13.2** Where and to the extent that the Regulations apply to the Works and

(a) the Engineer is appointed Planning Supervisor

and/or

(b) the Contractor is appointed Principal Contractor

then in taking any action as such they shall state in writing that the action is being taken under the Regulations.

13.3 (1) Any action under the Regulations taken by either the Planning Supervisor the Principal Contractor and in particular any alteration or amendment to the Health and Safety Plan shall be deemed to be an Engineer's instruction pursuant to Clause 2.3. Provided that the Contractor shall in no event be entitled to any additional payment and/or extension of time in respect of any such action to the extent that it results from any action lack of action or default on the part of the Contractor.

(2) If any such action of either the Planning Supervisor or the Principal Contractor could not in the Contractor's opinion reasonably have been foreseen by an experienced contractor the Contractor shall as early as practicable give written notice thereof to the Engineer.

ADDENDUM A

AVOIDANCE AND SETTLEMENT OF DISPUTES

Avoidance of disputes **A.1** In order to overcome where possible the causes of disputes and in those cases where disputes are likely still to arise to facilitate their clear definition and early resolution (whether by agreement or otherwise) the following procedure shall apply for the avoidance and settlement of disputes.

Matters of dissatisfaction **A.2** If at any time

(a) the Contractor is dissatisfied with any act or instruction of the Resident Engineer or any other person responsible to the Engineer or

(b) the Employer or the Contractor is dissatisfied with any decision opinion instruction direction certificate or valuation of the Engineer or with any other matter arising under or in connection with the Contract or the carrying out of the Works

the matter of dissatisfaction shall be referred to the Engineer who shall notify his written decision to the Employer and the Contractor within 14 days of the reference to him.

Disputes **A.3** The Employer and the Contractor agree that no matter shall constitute nor be said to give rise to a dispute unless and until in respect of that matter

(a) the time for the giving of a decision by the Engineer on a matter of dissatisfaction under Clause A2 has expired or the decision given is unacceptable or has not been implemented and in consequence the Employer or the Contractor has served on the other and on the Engineer a notice in writing (hereinafter called the Notice of Dispute) or

(b) an adjudicator has given a decision on a dispute under Clause A6 and the Employer or the Contractor is not giving effect to the decision and in consequence the other has served on him and the Engineer a Notice of Dispute

and the dispute shall be that stated in the Notice of Dispute. For the purposes of all matters arising under or in connection with the Contract or the carrying out of the Works the word 'dispute' shall be construed accordingly and shall include any difference.

A.4 (a) Notwithstanding the existence of a dispute following the service of a Notice under Clause A.3 and unless the Contract has already been determined or abandoned the Employer and the Contractor shall continue to perform their obligations.

(b) The Employer and the Contractor shall give effect forthwith to every decision of

(i) the Engineer on a matter of dissatisfaction given under Clause A.2 and

(ii) the adjudicator on a dispute given under Clause A.6

unless and until that decision is revised by agreement of the Employer and Contractor or pursuant to Addendum A.

Conciliation **A.5**

(a) The Employer or the Contractor may at any time before service of a Notice to Refer to arbitration under Clause A.9 by notice in writing seek the agreement of the other for the dispute to be considered under "The Institution of Civil Engineers' Conciliation Procedure (1999)" or any amendment or modification thereof being in force at the date of such notice.

(b) If the other party agrees to this procedure any recommendation of the conciliator shall be deemed to have been accepted as finally determining the dispute by agreement so that the matter is no longer in dispute unless a Notice of Adjudication under Clause A.6 or a Notice to Refer to arbitration under Clause A.9 has been served in respect of that dispute not later than 1 month after receipt of the recommendation by the dissenting party.

Adjudication **A.6**

(a) The Employer and the Contractor each has the right to refer any dispute as to a matter under the Contract for adjudication and either party may give notice in writing (hereinafter called the Notice of Adjudication) to the other at any time of his intention so to do. The adjudication shall be conducted under "The Institution of Civil Engineers' Adjudication Procedure (1997)" or any amendment or modification thereof being in force at the time of the said Notice.

(b) Unless the adjudicator has already been appointed he is to be appointed by a timetable with the object of securing his appointment and referral of the dispute to him within 7 days of such notice.

(c) The adjudicator shall reach a decision within 28 days of referral or such longer period as is agreed by the parties after the dispute has been referred.

(d) The adjudicator may extend the period of 28 days by up to 14 days with the consent of the party by whom the dispute was referred.

(e) The adjudicator shall act impartially.

(f) The adjudicator may take the initiative in ascertaining the facts and the law.

A.7 The decision of the adjudicator shall be binding until the dispute is finally determined by legal proceedings or by arbitration (if the contract provides for arbitration or the parties otherwise agree to arbitration) or by agreement.

A.8 The adjudicator is not liable for anything done or omitted in the discharge or purported discharge of his functions as adjudicator unless the act or omission is in bad faith and any employee or agent of the adjudicator is similarly not liable.

Arbitration **A.9**

(a) All disputes arising under or in connection with the Contract or the carrying out of the Works other than failure to give effect to a decision of an adjudicator shall be finally determined by reference to arbitration. The party seeking arbitration shall serve on the other party a notice in writing (called the Notice to Refer) to refer the dispute to arbitration.

(b) Where an adjudicator has given a decision under Clause A6 in respect of the particular dispute the Notice to Refer must be served within three months of the giving of the decision otherwise it shall be final as well as binding.

Appointment of arbitrator **A.10**

(a) The arbitrator shall be a person appointed by agreement of the parties.

President or Vice-President to act

(b) If the parties fail to appoint an arbitrator within one month of either party serving on the other party a notice in writing (hereinafter called the Notice to Concur) to concur in the appointment of an arbitrator the dispute shall be referred to a person to be appointed on the application of either party by the President for the time being of The Institution of Civil Engineers.

(c) If an arbitrator declines the appointment or after appointment is removed by order of a competent court or is incapable of acting or dies and the parties do not within one month of the vacancy arising fill the vacancy then either party may apply to the President for the time being of The Institution of Civil Engineers to appoint another arbitrator to fill the vacancy.

(d) In any case where the President for the time being of The Institution of Civil Engineers is not able to exercise the functions conferred on him by this Clause the said functions shall be exercised on his behalf by a Vice-President for the time being of the said Institution.

Arbitration–procedure and powers **A.11**

(a) Any reference to arbitration under this Clause shall be deemed to be a submission to arbitration within the meaning of the Arbitration Act 1996 or any statutory re-enactment or amendment thereof for the time being in force. The reference shall be conducted in accordance with the procedure set out in the Appendix to the Conditions of Contract or any amendment or modification thereof being in force at the time of the appointment of the arbitrator. Such arbitrator shall have full power to open up review and revise any decision opinion instruction direction certificate or valuation of the Engineer or an adjudicator.

(b) Neither party shall be limited in the arbitration to the evidence or arguments put to the Engineer or to any adjudicator pursuant to Clause A.2 or A.6 respectively.

(c) The award of the arbitrator shall be binding on all parties.

(d) Unless the parties otherwise agree in writing any reference to arbitration may proceed notwithstanding that the Works are not then complete or alleged to be complete.

Witness **A.12**

(a) No decision opinion instruction direction certificate or valuation given by the Engineer shall disqualify him from being called as a witness and giving evidence before a conciliator adjudicator or arbitrator on any matter whatsoever relevant to the dispute.

(b) All matters and information placed before a conciliator pursuant to a reference under Clause A.5 shall be deemed to be submitted to him without prejudice and the conciliator shall not be called as witness by the parties or anyone claiming through them in connection with any adjudication arbitration or other legal proceedings arising out of or connected with any matter so referred to him.

ICE Conditions of Contract for Minor Works

APPENDIX TO THE CONDITIONS OF CONTRACT
(to be prepared before tenders are invited and to be included with the documents supplied to prospective tenderers)

1. Short description of the work to be carried out under the Contract

....

....

....

2. The payment to be made under Article 2 of the Agreement in accordance with Clause 7 will be ascertained on the following basis.

(a) * Lump sum

(b) * Measure and value using a priced Bill of Quantities

(c) * Valuation based on a Schedule of Rates (with an indication in the Schedule of the approximate quantities of major items)

(d) * Valuation based on a Daywork Schedule

(e) * Cost plus (the cost is to be specifically defined in the Contract and will exclude off-site overheads and profit)

* *The alternatives not being used are to be deleted. Two or more bases for payment may be used on one Contract.*

3 Where a Bill of Quantities or a Schedule of Rates is provided the method of measurement used is

....

4 Name of Engineer (Clause 2. 1)

....

5 Starting date (if known) (Clause 4. 1)

....

6 Period for completion (Clause 4.2)

....

7 Period for completion of parts of the Works if applicable and details of the work to be carried out within each such part (Clause 4.2)

	Details of work	Period for completion
Part A	....	....
Part B	....	....
Part C	....	....

8 Liquidated damages (Clause 4.6)

....

9 Limit of liquidated damages (Clause 4.6)

....

10 Defects Correction Period (Clause 5. 1)

....

11 Rate of retention (Clause 7.3)

....

12 Limit of retention (Clause 7.3)

....

13 Minimum amount of interim certificate (Clause 7.3)

....

14 Bank whose base lending rate is to be used (Clause 7.8)

....

15 Insurance of the Works (Clause 10. 1)

Required/Not required [a]

16 Minimum amount of third party insurance (persons and property) (Clause 10.6)

....
for each and every occurrence

17 Name of the Planning Supervisor (Clause 13(1)(b))

....

Address ....

....

18 Name of the Principal Contractor (Clause 13(1)(b))

....

Address ....

....

19 The Arbitration Procedure to be used is (Clause A.11(a))

(a) "The Institution of Civil Engineers' Arbitration Procedure (1997)"[a]

or

(b) "The Construction Industry Model Arbitration Rules "[a]

If no deletion is made "The Institution of Civil Engineers' Arbitration Procedure (1997)" shall be used.

[a] *Delete as appropriate.*

ICE Conditions of Contract for Minor Works

AGREEMENT

THIS AGREEMENT is made the .... day of ....20 ......

between ....

of (or whose registered office is at)..

....

(called the 'Employer') of the one part

and..

of (or whose registered office is at)..

....

(called the 'Contractor') of the other part

WHEREAS the Employer wishes to have carried out the following ....

....

....

and has accepted a Tender by the Contractor for the same

NOW IT IS HEREBY AGREED AS FOLLOWS:

Article 1
The Contractor will subject to the Conditions of Contract perform and complete the Works.

Article 2
The Employer will pay the Contractor such sum or sums as shall become payable under the Contract and in accordance with the Conditions of Contract.

Article 3
The following documents shall be deemed to form and be read and construed as part of this agreement namely:-

The Contractor's Tender (excluding any general or printed terms contained or referred to therein unless expressly agreed in writing to be incorporated in the Contract)

The Conditions of Contract

The Appendix to the Conditions of Contract

The Drawings. Reference numbers ....

....

....

The Specification. Reference ....

The priced Bill of Quantities*

The Schedule of Rates*

The Daywork Schedules*

The following extra documents *

from .... to ....dated ....

from .... to ....dated ....

from .... to ....dated ....

from .... to ....dated........................

SIGNED on behalf of the said .... Ltd/plc (the Employer)

Signature...

Position...

In the presence of..

SIGNED on behalf of the said .... Ltd/plc (the Contractor)

Signature…..

Position. ....

In the presence of..

* *Delete if not applicable*

ICE Design and Construct Conditions of Contract

Second Edition

GUIDANCE NOTES

The Institution of Civil Engineers
Association of Consulting Engineers
Civil Engineering Contractors Association

Published by Thomas Telford Publishing, Thomas Telford Limited, 1 Heron Quay, London E14 4JD (Telephone +44 (0)207 665 2462 Fax +44 (0)207 538 5746
E-mail customerservices@thomastelford.com) on behalf of:

The Chief Executive and Secretary
The Institution of Civil Engineers
One Great George Street
London SW1P 3AA

The Chief Executive
Association of Consulting Engineers
Alliance House
12 Caxton Street
London SW1H 0QL

The Director
Civil Engineering Contractors Association
Construction House
56-64 Leonard Street
London EC2A 4JX

9 8 7 6 5 4 3 2 1

ISBN 0 7277 3024 X

A catalogue record for this book is available from the British Library

Printed and bound in Great Britain by
Selwood Printing Ltd., Burgess Hill, West Sussex

CONTENTS

1 INTRODUCTION

After the publication in June 1973 of the Fifth Edition of the ICE Conditions of Contract a permanent joint committee – the Conditions of Contract Standing Joint Committee (CCSJC) – was set up to keep the use of standard forms of contract under review, to consider any suggestions for amendment and also to undertake the drafting of new standard forms.

The composition of the CCSJC is as follows:

An independent chairman appointed by the ICE.

Three members appointed by the ACE to represent Consultants.

Three members appointed by the CECA to represent Contractors.

Three members appointed by the ICE on the advice of client bodies to represent Clients.

One co-opted member (a Fellow or Member of the ICE with an academic or legal background) appointed by the ICE.

The terms of reference of the CCSJC exclude the provision of any legal interpretation.

Traditionally the ICE Conditions of Contract for Works of Civil Engineering Construction have been drafted for use in connection with projects for which the design of the Works is provided by or on behalf of the Employer and the construction is carried out by the Contractor under the supervision of the Engineer, although under the Measurement Version and other forms provision is made for Contractor-design of specified parts of the Works. Payment has usually, but not necessarily, been based on a system of admeasurement (or "measure and value") although other modes of payment are also contemplated.

The ICE Design and Construct Conditions of Contract depart radically from this concept with the Contractor responsible for all aspects of design and construction, including any design originally provided by or on behalf of the Employer. The Form of Tender provides for payment on a lump sum basis, but other forms of payment or combinations of forms may be used.

These Conditions of Contract are therefore not a variant of the Measurement Version but an entirely separate set of Conditions based on very different premises. The layout, Clause numbering and in some cases the wording are similar to those of the Measurement Version but the import of the Clauses is in many cases very different.

These Guidance Notes have been prepared by the CCSJC specifically to assist users of the ICE Design and Construct Conditions of Contract in the preparation of contract documents and the carrying out of contract works. They do not purport to provide legal interpretation but they do represent the unanimous view of the CCSJC on what constitutes good practice in the conduct of civil engineering projects of this type.

The Contract is drafted bearing in mind the benefits of team working and current procurement initiatives. If its procedures are followed the parties to the Contract will be provided with an "early warning" of circumstances that may give rise to additional costs or delay in a stage or in completion of the Contract.

The Contract is also drafted with the intention of minimising the incidence of potential disputes and resolving those that may arise in a speedy and non-confrontational manner. The Contract provides the choice of conciliation, adjudication and arbitration to facilitate the resolution of disputes.

The Housing Grants, Construction and Regeneration Act 1996 came into force on 1 May 1998 for construction contracts as defined in the Act. The Contract is drafted to comply with this legislation with the intention that the Statutory "Scheme for Construction Contracts" should not apply (see in particular Clauses 60 and 66). Clause 67 deals with the application of the legislation to Scotland and the Construction Contracts (Northern Ireland) Order in Northern Ireland.

Compared with the previous edition, several substantial changes and a number of minor ones have been made and a certain amount of editing has been carried out to improve the clarity of the document.

It should be noted that these Guidance Notes should not and do not form part of the Contract.

2 OVERALL CONCEPT

2.1 Contractor's overall responsibility

The basis of the ICE Design and Construct Conditions of Contract is that of an Employer, who sets out in the Employer's Requirements the design brief and other matters relating to design, construction and performance which must be included or adhered to in carrying out the Works, and a Contractor who then develops the Employer's Requirements and designs and constructs the Works (Clause 8).

2.2 Employer's Requirements

The success of the Contract depends upon a clear understanding by the Contractor of the Employer's Requirements (Clause 1(1)(e)). These are initially set out in the document issued to tenderers and may be modified prior to the award of the Contract. Their scope may vary from a statement of the overall requirement to a far more detailed description including outline design, specifications, standards of performance, general appearance and other matters.

Provision is made in Clause 51 for alteration to the Employer's Requirements by the Employer's Representative during the currency of the Contract but, because of the nature of the Contract, such alterations should be kept to the absolute minimum. Provision is also made for the Contractor to propose such alterations with the possibility of sharing any gains or losses arising.

2.3 Contractor's Submission

In response to an invitation to tender, contractors are required to submit tenders setting out in a series of documents their scheme for the Works (Clause 1(1)(f)). These documents are required to be listed in the Form of Tender.

It must be the aim of the parties in discussions, possibly during the tender period, and in any event before the award of the Contract, so to amend the Employer's Requirements and the Contractor's Submission that they set out an agreed scheme and specification for the Works.

In interpreting the final Contract the Employer's Requirements take precedence over the Contractor's Submission (Clause 5(1)).

2.4 Employer's Representative

Once the Contract has been awarded it is necessary for the Contractor to have ready access to a person with appropriate authority on the Employer's side so as to enable the Contractor to finalise and detail the design without undue delay and to deal with any problems which may arise during construction and which require approval. This person is the Employer's Representative (Clauses 1(1)(c) and 2).

Any alterations to the Employer's Requirements required by the Employer during the currency of the Contract (after consultation with the Contractor's Representative) are to be conveyed to the Contractor by the Employer's Representative but the development of such alterations into design and construction changes remains the Contractor's responsibility (Clauses 8(1)–(2) and 51).

The full range of responsibility and authority given to the Employer's Representative under the Contract is set out in Section 7 of these Guidance Notes.

It is important to understand that the Employer's Representative does not have the same role as the Engineer under the other ICE forms of contract. The Contractor is responsible for ensuring that work carried out is acceptable under the Contract; this duty is entirely his, including using such Quality Assurance procedures as the Contract may require (Clause 8(3)). Any inspections which the Employer's Representative may make are solely to verify that the Works being constructed are in accordance with the Contract and, in particular, meet the Employer's Requirements.

2.5 Payment The Form of Tender contains an offer to complete the Works for "the sum of .... or such other sum as may be ascertained in accordance with the said Conditions of Contract."

The Conditions of Contract permit changes to the sum and possible additional payments in accordance with the terms of Clauses 5, 6, 12, 13, 18, 31, 36, 40, 42, 51, 52 and 58.

Other methods of valuation could, however, be adopted by amendment of the Form of Tender and addition of details of the means to be used. Clause 55 allows for measurement where required.

Clause 60(2)(a) provides for use of a payment schedule for interim payments. This could be called for or submitted with the tender. Clause 60(2)(b) covers the situation where no such schedule exists. A schedule or other means of assessing payments may be included in the Tender Appendix – Part 2.

Consideration must be given by the Employer, pre-tender, with regard to the inclusion of a special condition relating to Contract Price Fluctuation and, if one is to be included, the method of calculation. Standard clauses are bound into the Conditions of Contract for use if required. See also Appendix 2 to these notes for guidance.

3 CHOICE OF DESIGN AND CONSTRUCT CONTRACT

A prospective Employer should consider using the ICE Design and Construct Conditions of Contract where:

(1) the Employer's Requirements can be clearly set out and understood, with little risk of changes being required during the period of the contract.

(2) circumstances suggest that the use of this form of Contract may expedite successful completion of the Works, having regard to such matters as the availability of suitable contractors, programme restraints and the desirability of a simplified management structure and

(3) sufficient information is available to tenderers to allow certainty of pricing.

Employers should recognise that the Contractor will be responsible for determining all matters not fixed by the Employer's Requirements and that any alterations to these insisted on by the Employer may lead to extra payments and delay completion.

The overall timescale for a project may, in some situations, be reduced by using the single provider of both design and construction rather than the traditional sequential process.

These conditions should not, however, be used where the Employer seeks to compress the design and construction processes to meet a restricted period for completion, since this would place an unfair burden of risk on to the Contractor. The compression of a construction programme is better achieved by the adoption of conditions under which the Employer himself provides or controls the design process.

4 TENDERING PROCEDURE

4.1 Tenderers Tenderers should be drawn from firms with the necessary experience of the type of project that the Employer wishes to procure. They should have access to the appropriate design expertise and have knowledge of the materials required for, and the methods of construction, appropriate to such a project.

Consideration should be given to the need for comprehensive pre-qualification of tenderers.

4.2 Number of tenderers Because of the high cost and extensive work involved in tendering for this type of work good practice dictates that not more than four, and it is recommended that only three, firms or consortia should be invited to tender for a design and construct project. Consideration could also be given to payment of a sum to all tenderers who submit a complete tender.

There may be over-riding legislative requirements that modify the above recommendations, these include UK Contracts Regulations or EC Public Procurement Directives.

4.3 Tender period It is an essential feature of this type of contract that the contractor takes responsibility for the design. Tenderers must thus be given adequate time to consider and prepare outline designs so as to allow realistic pricing. Since contractors will, under Clause 8(2), take over responsibility for any design identified in the Employer's Requirements, the time allowed must include for any necessary checks on such designs.

4.4 Information to tenderers In addition to information included in the Appendix to the Form of Tender, tenderers will need to be given:

(1) The Employer's Requirements (see also section 4.5 of these notes) which will set out:

(a) the Employer's design brief and also details of matters which are not to be left to the Contractor regarding the general layout, design requirements, appearance, plant and equipment to be installed, performance requirements, offices and/or other accommodation required, accesses, etc., and any limitations on Site availability (Clause 42(1))

(b) any performance specifications for the Permanent Works and/or any general specifications stipulated by the Employer

(c) any items and activities for which the Contractor is to establish and implement an approved quality assurance system for the mutual benefit of the Employer and the Contractor (Clause 8(3))

(d) any particular matters affecting the safety of adjacent structures or existing works which may affect the design or methods of construction and

(e) the length of the Defects Correction Period (Clause 1 (1)(p)) or any other special Defects Correction liabilities.

(2) Details of any particular information that is to be included in or with the Contractor's Submission, including details of insurance cover required, details of items of plant and any special materials proposed for use and details of any special types of design or construction to be used.

(3) Details of site surveys, ground investigation reports, hydrological surveys and details of pipes and cables in, on or over the ground that have been obtained by or on behalf of the Employer and which are relevant to the Works (Clause 11 (1)).

(4) The envisaged tender acceptance date and Works Commencement Date (Clause 41(1)), with a clear statement as to whether these dates are definite or are given only for guidance.

It should be noted that, due to the nature of this type of contract, no provision is made for the use of nominated sub-contractors.

4.5 Employer's Requirements

The Employer's Requirements will be set out in a number of documents and drawings intended to define what the Employer wants to achieve from the Contract. The Employer's Requirements are perhaps the most important documents in the Contract. They must be prepared with great care and be clear and unambiguous; any alterations to these requirements after award will probably result in cost to the Employer. It must be remembered that, after award, the Contractor takes over responsibility for all design (except for any items specifically defined in the Employer's Requirements) and his obligation then is only to produce designs and construction to meet the Employer's Requirements and no more. The Employer's Requirements should thus cover all those matters essential to the Employer but leave to the Contractor all other non - essential matters.

It is essential in the Employer's Requirements to differentiate between those matters that are mandatory requirements and those that are intended as guidance or to be included in the design if needed.

4.6 Examination of tenderers' submissions

The Employer (using any necessary expert advice) will need to satisfy himself that tenderers' submissions for the overall layout and design and construction of the project are sound in concept and would fulfil his needs and requirements (whether expressly stated by him or not). He should also satisfy himself that the construction and payment programmes, together with any particular circumstances arising and risks involved, are acceptable to him.

Tenderers' submissions are likely to contain matters that need to be clarified and possibly adjusted. Agreements reached at any resulting negotiations and meetings with each tenderer will need to be recorded accurately, especially in respect of any matters where a Contractor's Submission differs from the Employer's Requirements as issued to tenderers, in order that any resulting Contract will be unambiguous. The Employer may need to consider whether it is appropriate to alter the Employer's Requirements prior to acceptance.

4.7 Performance requirements

The Employer's Requirements will normally include any performance requirements to be met. These may be either as fixed targets to be warranted or criteria to be measured. For some types of work it may be appropriate to specify performance tests to demonstrate compliance. These may be carried out prior to completion but it is often necessary to allow plant to operate for a while before testing. Any tests required should be set out in the Employer's Requirements stating when and who will carry out the tests, any equipment, materials and services to be provided and the timing and range of operations to be covered.

Consideration should be given to the need for performance related damages to be paid if targets are not met or perhaps a sliding scale of damages and bonuses to apply to measured criteria.

5 AWARD OF THE CONTRACT

After the Employer has made his choice of tenderer, the Employer will probably proceed by issuing a Letter of Acceptance to the successful tenderer (who thereupon becomes the Contractor). The Letter of Acceptance should set out the documents, together with any agreed amendments thereto and any other agreements made subsequent to the submission of the tender that will form the Contract. These will include the Contractor's Submission (the tenderer's submission as amended) and the updated Employer's Requirements.

If required the Contractor will then enter into and execute a Contract Agreement as set out in Clause 9.

6 OPERATION OF THE CONTRACT

6.1 Appointment of Employer's and Contractor's Representatives Operation of the Contract relies upon a close working relationship between the Employer and the Contractor (through the Employer's Representative nominated under Clauses 1(1)(c) and 2 and the Contractor's Representative nominated under Clauses 1(1)(d) and 15(2)) to ensure that the Employer's Requirements are properly understood and carried out. It is important that both of these nominations are made at the start of the Contract.

6.2 Employer's Representative The Employer's Representative is the agent of the Employer for the operation of the Contract and is to be a named individual (Clauses 1 (1)(c) and 2(2)).

Although not necessarily required to be an engineer, the Employer's Representative must be fully conversant with the Employer's Requirements and the essential thinking behind them and of conditions on the Site so as to be able to:

(1) ensure prompt response to the Contractor's requests for consent to the general arrangements and to detailed designs in order to meet the Employer's Requirements (Clause 6(2))

(2) provide any necessary information in response to the Contractor's enquiries, and

(3) know when changes need to be made in the Employer's Requirements by instructing alterations under Clause 51 (1).

6.3 Contractor's Representative The Contractor is required to appoint a representative to exercise overall superintendence of the Works and to have authority to act and commit the Contractor as if he were the Contractor for the purposes of the Contract (Clauses 1(1)(d) and 15(1)-(2)).

The Contractor's Representative or a competent and authorised agent approved in writing by the Employer's Representative is, throughout the period of construction of the Works, to give his whole time to superintendence of the work on Site (Clause 15(3)). His responsibilities will normally include matters relating to safety during construction. See notes for Clauses 19 and 71 in Section 8.

6.4 Information release schedule To facilitate and expedite co-operation between the Employer, the Employer's Representative, the Employer's other advisers and suppliers and the Contractor it is recommended that an information release schedule should be produced by the Contractor early in the course of the Contract. This will set out the dates when information is required from the Employer and designs etc. are required from the Contractor and others so that design and construction can progress in conformity with the programme provided under Clause 14.

Submissions of designs and drawings are required from the Contractor under Clause 6. These are for review by the Employer's Representative to check compliance with the Employer's Requirements. The Contractor's Representative and the Employer's Representative should agree reasonable requirements for such submissions, including timing, if these have not already been set out in the Employer's Requirements.

Communication will normally be between the Employer's Representative and the Contractor's Representative and their respective Site staffs, and requests and data from other participants should be routed accordingly.

6.5 Contractor's design responsibility

The Contractor is required to list in the Appendix to the Form of Tender the names of the designers who will be responsible for the various parts of the Works and to obtain the consent of the Employer to any change in designer (Clause 4(2)(a)).

In some cases, where the Employer has appointed a design team to prepare his initial designs, it may be appropriate to have the same team continue with the detailed designs for the Contractor. This may involve a novation of the design contract to the Contractor. Any such arrangement should be set out in the Employer's Requirements or be agreed between the parties at an early stage.

The Contractor's design responsibility is set out particularly in Clauses 6 and 8 and includes the specification and selection of materials and plant and the safety of design.

There may be some matters or items for which the Employer retains design responsibility. These need to be identified in the Employer's Requirements. Such matters might, for example, include use of a specific manufacturer's plant or a design for connection to existing works.

Where any part of the Works has been designed by or on behalf of the Employer, but design responsibility is to pass to the Contractor then that design must be identified in the Employer's Requirements. The Contractor is required to check that design. This check takes place after award of the Contract and allows for any essential changes to the existing designs necessary to meet the Employer's Requirements as then set out. Following this check the Contractor must obtain the approval of the Employer's Representative to any design modifications that he considers necessary. Notwithstanding any checks carried out when tendering, the Contractor then accepts responsibility for the combined design. Any queries raised during tendering or in any negotiations should have been settled prior to award by alterations to the Contractor's Submission or the Employer's Requirements.

Where checks on designs are required by any Act of Parliament or Byelaw the Contractor is responsible for arranging and paying for such checks (Clause 8(4)).

Where the Contractor considers it necessary or desirable to carry out further site investigation he is responsible for so doing subject to permission from the Employer's Representative (Clause 18(1)).

The Contractor is required to obtain the consent of the Employer's Representative to his designs and to his general arrangement and such other drawings as are necessary to demonstrate that the Works comply with the Employer's Requirements (Clause 6(2)).

6.6 Professional indemnity insurance

The Contractor is not required by the Contract to carry professional indemnity insurance against negligence in the design. He could, however, be liable to the Employer should his designer fail to exercise reasonable skill, care and diligence (Clause 8(2)(a)) resulting in damage to the Employer. The Contractor will need to bear this in mind in determining both the level of design checks and the procedures for carrying them out. Such damage may not be covered by insurances taken out under Clauses 21 and 23.

The maintenance of professional indemnity insurance in respect of design may be included in the Employer's Requirements but the following should be taken into account:

(1) Professional indemnity insurances are annual policies that have to be renewed every year.

(2) Possession of such a policy is no guarantee that it can be renewed.

(3) The cover under such a policy applies only to claims made in that year and not to those arising from negligence in the year that gives rise to a claim in a subsequent year.

6.7 Checking of design, materials and workmanship by the Contractor

Responsibility for ensuring that the work carried out is acceptable under the Contract lies entirely with the Contractor (Clause 36(1)).

The Contractor is required by Clauses 36(3) and 36(6) to institute a system of checks on his design and setting out of the Works and for testing all materials and workmanship. In addition the Contractor may be required to institute a quality assurance system (Clause 8(3)).

All such systems and procedures require the approval of the Employer's Representative.

6.8 Resolving problems

Should information be required or problems arise or should the Contractor wish to make any claim then he should give early notice to the Employer's Representative so that the matters can be dealt with expeditiously. Times for submission of such notices are given in many clauses to prompt early resolution. The representatives of the parties should have authority to resolve most matters likely to arise between themselves and under the arrangements set out in the various clauses. If either party is dissatisfied with the outcome then the matters can be referred to the Employer's Representative for his decision and, if a dispute is then established, to resolution under Clause 66. See further comments in Part 8 of these notes.

7 EMPLOYER'S REPRESENTATIVE

7.1 Introduction

The ICE Design and Construct Conditions of Contract introduce the Employer's Representative, whose functions are defined. However, as such an individual does not exist in other sets of ICE Conditions of Contract and as his function in the Design and Construct Contract is fundamental to the proper operation of that Contract the following points need to be stressed in addition to those set out in Section 6.2 of these Guidance Notes.

7.2 Responsibility in respect of design and construction

The Employer's Representative has no responsibility for providing designs. He is required to review designs produced by the Contractor or his designers (Clause 6). The main purpose of this review is to ensure that the designs will meet the Employer's Requirements and any statutory or other regulations that may apply. A further purpose will be to check that, taken together, the Contractor's designs and the Employer's Requirements will fulfil the purposes of the Works. Once consented to, the Contractor may not change his designs without further consideration by the Employer's Representative.

As is usual, the Contractor has responsibility for construction of the Works but the Employer's Representative will need to consider the construction methods proposed to ensure that the Employer's Requirements are met and that the Works can be constructed in accordance with any specific requirements or restrictions imposed by the Contract. See Clause 6(4). While the Contractor remains responsible for safety matters on site, the Employer's Representative retains the normal responsibilities for his staff and to warn of any potentially unsafe practices that he may observe.

It is important to remember that it is the Contractor's responsibility to design and construct the Works to meet the Employer's Requirements. As a result, the Employer's Representative may not have the detailed knowledge of the project that would be available to the Engineer in other ICE forms. The role of the Employer's Representative thus differs significantly from that of the Engineer in other forms of ICE contracts, his task being mainly to see that the Employer's Requirements are met.

If the Employer's Representative considers that the Employer's Requirements will not be met or that the designs or methods of construction will not comply with the Contract then he may require the Contractor to make further submissions (Clause 6(2)). Further, under Clause 13, the Employer's Representative may issue instructions to ensure compliance with the Contract. Such instructions should be given only by the Employer's Representative and only after consultation with the Contractor. This provision is not intended to allow the Employer's Representative to interfere with the Contractor's operations but to prevent matters going irretrievably wrong by keeping the Contractor to his contractual obligations. Clearly the Employer's Representative must behave impartially in reaching his decisions on these matters. If the instructions cause delay that is not due to the Contractor's default, he may be entitled to extension of time or to recover costs. If such instructions involve alteration to the Employers Requirements then they are deemed given under Clause 51 and the Contract Sum adjusted accordingly.

The Employer's Representative has the right to inspect the work to satisfy himself that the requirements of the Contract are being met.

7.3 Delegation by the Employer's Representative

The Employer's Representative may delegate any of his functions except those under Clauses 12(6), 13, 15(2)(b), 44, 46(3), 48, 60(4), 61, 65 and 66 to others, including assistants (Clause 2(3)-(4)), whose names and functions are to be notified to the Contractor. If appropriate, an assistant may be appointed specifically to watch the construction of the Works. Such an appointment would in no way relieve the Contractor of any of his obligations under the Contract (Clause 1(7)).

The Employer's Representative can take advice from appropriate experts and may delegate part or parts of his responsibilities to such experts under the terms and limitations of the Contract.

7.4 Responsibilities of the Employer's Representative in respect of planning, programme and alterations

(1) The Employer's Representative must be able readily and rapidly to respond to queries from the Contractor's Representative, especially during the planning and design stages (Clauses 6(1)).

(2) The Employer's Representative should be conversant with the conditions of the Site as they relate both to access and construction, including what is known of the ground conditions and underlying strata (Clauses 12(4)-(6), 40(1) and 42).

(3) The Contractor is required by the Contract to obtain the consent of the Employer's Representative to general arrangements and detailed designs relating to specific matters in the Employer's Requirements. The Employer's Representative must be ready to deal promptly with such applications and must also to be ready to accede to proper requests by the Contractor for permission to carry out any necessary additional ground investigations (Clauses 6 and 18).

(4) The Employer's Representative should ensure, on the Employer's behalf, that the required parts of the Site are made available to the Contractor in accordance with Clause 42.

(5) The Employer's Representative is required to consider the Contractor's programme or require its revision under Clauses 14(1) and (2) and 14(4), to authorise times of working under Clause 45, and if necessary to require that the Contractor expedite progress under Clause 46.

The Employer's Representative should be well acquainted with the progress of the Works on Site so as to be able to take such action under Clauses 14(4), 44 and 46 as may be necessary.

(6) The power of the Employer's Representative to order alterations under Clause 51 is limited to altering the Employer's Requirements. See also the comments in Section 8 of these notes.

The ordering of work on a daywork basis is covered by Clause 52(4) and expenditure under Provisional Sum and Prime Cost Items is covered by Clause 58.

7.5 Responsibilities of the Employer's Representative for valuation and certification

The Employer's Representative is required by the Contract:

(1) where necessary, to examine work and measure it before it is covered up as provided by Clauses 38 and 56(1)

(2) to negotiate and agree, where possible, the value of alterations under Clause 52

(3) to approve transfer of ownership of goods to be vested in the Employer under Clause 54

(4) to certify payments due to the Contractor under Clause 60

(5) to certify the Substantial Completion of the various parts of the Works under Clauses 48(2) and 48(4) and the completion of outstanding work during the Defects Correction Period under Clause 61(1).

7.6 Responsibilities of the Employer's Representative for making decisions

The Employer's Representative acts as the agent of the Employer to supervise the design and construction of the Works and to satisfy himself that the Works are being constructed in accordance with the Contract and will fulfil the Employer's Requirements. Of no less importance are those functions which require the Employer's Representative to make decisions regarding any alterations to the Employer's Requirements that need to be made in the course of construction or which arise from the allocation of risks. Clause 2(6) expressly requires the Employer's Representative to act impartially in respect of the main certifying, valuing and decision making functions. While recognising that he represents the Employer's interests, the Employer's Representative should also act reasonably and fairly in accordance with the Contract in all his functions with a view to resolving matters promptly and properly.

The Employer's Representative has the duty to exercise his professional judgement, honestly and impartially to decide such matters as unforeseen conditions (Clause 12), extensions of time (Clause 44), the valuation of alterations to the Employer's Requirements (Clause 52) and to decide formally any issues where either the Employer or Contractor are dissatisfied (Clause 66). While the Employer's Representative will make many decisions during a contract, it should be remembered that any of these can be disputed by either party and referred to final determination under the provisions of Clause 66.

8 NOTES ON SPECIFIC CLAUSES

Clause 1(1)(a): Employer

The Employer has a number of duties to perform under the Contract.

If a Contract Agreement (Clause 9) is required by the Employer, he is responsible for its preparation. The Employer also has responsibility for approving the Contractor's performance security (Clause 10(1)). For his own protection he should also check the terms and coverage of the Contractor's insurance policies and verify that premiums have been paid (Clauses 25(1) and (3)).

The Employer is responsible for ensuring that the Contractor will be given possession of and access to the Site to enable him to commence and proceed with the Works. Any restrictions must be set out in the contract documents. The Employer is also responsible for ensuring that payments are made to the Contractor by the final date for payment following certification by the Employer's Representative.

The Employer has the power, in certain circumstances, to terminate the employment of the Contractor.

Clause 1(7): Consent and approval

The responsibility for compliance with the Contract remains with the Contractor despite any approval or failure to approve or objection to any of the Contractor's proposals or submissions by the Employer's Representative.

Clause 2: Duties and authority of Employer's Representative

The Employer's Representative is appointed by the Employer to monitor the Contract and to that end has conferred on him by both Parties the powers and duties vested in him under the Contract.

It is in the interests of both the Employer and the Contractor that any queries or problems that arise during construction should be resolved expeditiously. To this end the Contract requires that the Employer's Representative's function shall be exercised by a single named person. See also Section 7.

Clause 3(2): Rights of Third Parties

By this Clause the acquisition of rights by third parties is excluded. Should the parties to the Contract wish to grant specific rights to particular third parties they may do so by means of additional suitable clauses included under special conditions.

Clause 4: Sub-contracting

Clause 4 recognises the practice in the construction industry of the use of sub-contractors and self-employed labour. In general the Contractor should be free to choose his own sub-contractors.

The Contractor's designer is to be named in Part 2 of the Appendix to the Form of Tender. Any proposed change of designer is subject to consent by the Employer.

There are two aspects to Clause 4(2). The first allows the Employer, at the tender stage, to advise the Contractor via the Appendix to the Form of Tender of those aspects of the work which the Contractor must not sub-contract to others without the specific authority of the Employer.

The second aspect requires the Contractor to advise the Employer's Representative of the name and address of a proposed sub-contractor at least 14 days in advance of that sub-contractor's entry on to the Site. The Contractor should take care not to create a binding obligation to his sub-contractor until he is sure the Employer's Representative will not object.

Any objections to a specific sub-contractor must be made within 7 days of receiving the notification from the Contractor. It is essential that the Employer's Representative not only gives his reasons for objection but also that these must be in writing. The reasons must be cogent and of substance, since the objection may be open to challenge. Caution must be exercised in the manner of objection to a specific sub-contractor so that legislation on procurement procedures is not infringed. If a specific sub-contractor is objected to without good reason it is possible the Contractor could be involved in additional costs for which he may be entitled to reimbursement under the Contract.

Under Clause 4(4) the Contractor is fully responsible for all sub-contracted work and for the acts, defaults and neglects of any sub-contractor or his employees. The Employer's Representative's acceptance of a sub-contractor in no way relieves the Contractor of this responsibility.

The provisions of Clause 4(5) (in respect of sub-contracting firms) and Clause 16 (in respect of individuals) give the Employer's Representative power to require the removal from Site of sub-contractors or other persons in the circumstances described.

Clauses containing provisions that could relate to design work that may be carried out by a sub-contractor are as follows:

> Clause 4(2) requires notification of any change to the Contractor's designer.
>
> Clause 4(5) covers the removal from the Works or their design of any sub-contractor in the circumstances described.
>
> Clause 6(2) and 6(3) cover the submission of designs and drawings.
>
> Clause 7(1) deals with the question of the copyright of drawings, specifications and other documents.
>
> Clause 8(2) deals with the question of design responsibility.
>
> Clause 8(3) refers to quality assurance requirements.
>
> Clause 16 contains provisions similar to those in Clause 4(5) and covers the dismissal from the Works of individuals in the circumstances described.
>
> Clause 48(3) covers the provision of operation and maintenance manuals.

Should the Employer wish to specify the use of a particular supplier or sub-contractor then this must be set out in the Employer's Requirements.

Clause 6: Further information

The Employer's Representative is responsible for providing any further information to which the Contractor is entitled under the Contract.

The Contractor is required to provide to the Employer's Representative designs and drawings to demonstrate that the proposed Works will comply with the Employer's Requirements. The Employer's Representative may ask for re-submissions until he is satisfied that the designs will comply and he can give his consent to them. Construction is not to commence until the designs and drawings have been consented to. Any subsequent changes are subject to further submissions and consent.

The Employer's Representative may also request information on methods of construction to demonstrate that the Works can be built in accordance with the Contract and without detriment to the permanent Works.

The information to be provided by the Contractor is that reasonably required by the Employer's Representative for good contractual reasons. He is **not** entitled to request details of everything.

Clause 8(2): Contractor's design responsibility

This Contract differs radically from other ICE forms in that it requires the Contractor to be fully responsible for design as well as construction. This includes taking over responsibility for any designs provided as part of the Employer's Requirements. See also Section 6.5 of these notes.

Clause 10: Performance Security

Various choices are available for providing security to the Employer against the possibility of a contractor's default. Where a contractor is financially substantial, a bond may be considered unnecessary or, if part of a large group, a Parent Company guarantee may be sufficient. The Form of Default Bond enclosed with the Conditions of Contract is suggested for use where the Employer requires an independent security. The bond is intended to provide the extra funding necessary to allow the Employer to complete the works following a default. The cost of providing bonds can be high and the amount of security required should be chosen to represent the likely extra cost involved. The contract limits the amount to 10% of the value of the Works but for many situations this may be too high.

The Form of Tender Appendix - Part 1 will indicate whether a bond is required and the amount. The Contractor will normally be required to meet the cost of providing this. For a Guidance Note on the Performance Bond see Appendix 1.

Clause 11: Provision and interpretation of information

Employers seek certainty of expenditure on a project and to assist in this objective it is beneficial to both Employer and Contractor for all information held to be provided at tender stage. The effect of Clause 11(1) is that information obtained by or on behalf of the Employer relevant to the Works shall only be taken into account to the extent that it was made available to the Contractor before the submission of his tender. The interpretation of all such information remains the Contractor's responsibility. If the Employer fails to make available such relevant information at the proper time he cannot later argue that it ought to have been known to the Contractor. The Contractor would, however, always be expected to have obtained for himself any information that was otherwise readily available and if possible to have visited the Site (Clauses 11(2) and 11(3)). The Contractor is deemed to have made his submission on the basis set out in sub-clauses 11(1) and (2).

Should the Employer produce fresh information after the Contract has been awarded, and such information then requires some change in the Contractor's Submission, his design or his method of construction, then that may be an alteration under Clause 51.

Clause 12: General — ***Adverse physical conditions and artificial obstructions***

It is impossible for either of the contracting parties to foresee every circumstance that may affect the design or construction of the Works. It is, however, an essential feature of the various forms of ICE Conditions of Contract that the Contractor accepts full responsibility for physical conditions or artificial obstructions that could reasonably have been foreseen by an experienced contractor. It is the Contractor's obligation to design and construct the Works and he must therefore propose any measures to deal with the conditions or obstructions encountered.

The Contractor is responsible for assessment of the effect that the circumstances encountered may have on his design and on construction but the consequences of any unforeseeable risks have to be dealt with separately in accordance with the terms of the Contract. Such circumstances may entitle the Contractor to additional payment or extra time.

It is clearly in the Employer's own interests to have an adequate Ground Investigation carried out before tenders are returned.

Clause 12: Adverse physical conditions and artificial obstructions

Clause 12(1) requires the Contractor to give written notification to the Employer's Representative as early as is practicable of any physical condition or artificial obstruction encountered which the Contractor considers could not reasonably have been foreseen whether or not there is any intention to claim additional payment or time. Such notification must give details of any anticipated effects, the measures being taken or proposed, including any alternative designs, and their estimated cost and any consequential delay and/or interference with the construction of the Works.

Notification of the intention to claim additional payment or an extension of time is still required and can be made at the same time or as soon as is reasonable thereafter (Clause 12(2)).

Following receipt of written notification, the options open to the Employer's Representative under Clause 12(4) include requiring the Contractor to investigate and report on any alternative measures that may be available for dealing with the situation.

Clause 13: Instructions of the Employer's Representative

This clause allows the Employer's Representative himself (not a delegate) to issue instructions if he considers that the Contractor's conduct will not comply with the Contract. The Employer's Representative must both consult with the Contractor before issuing instructions and give reasons for his opinions. See also Section 7.2 of these notes.

The Contractor must comply with any instructions under this clause but may recover time or costs if any subsequent delays are not due to his own default. If the instructions alter the Employer's Requirements, they are deemed to have been given under Clause 51.

Clause 14: Programme to be furnished

Clause 14 calls for the submission of a programme and general method statements for construction within 21 days after the award of the Contract (Clause 14(1)). Such a programme and method statements are intended as a management tool to indicate to the Employer's Representative when specific areas of the Site and/or additional information will be required by the Contractor and to enable an assessment to be made of any failure by the Contractor to maintain the required rate of progress on the Contract (Clause 46(1)). The Employer's Representative may, within 14 days, reject the programme giving his reasons or ask for further information to clarify.

In respect of the programme the Contractor then has 14 days in which to submit a revised programme or provide the further information requested. When the Contractor submits a revised programme a further 14 day period is allowed for rejection.

There is no specific requirement for acceptance by the Employer's Representative of the programme. Responsibility for proceeding with the Works in compliance with the Contract remains with the Contractor and he may proceed unless an objection is raised by the Employer's Representative.

If the Employer's Representative considers that actual progress is not conforming to the programme he may require the Contractor to provide a revised programme to show compliance with the Contract.

Clause 19: Safety and security

Liability for safety is imposed by the Health and Safety at Work etc. Act 1974 and other statutes. It would therefore be inappropriate and could be misleading for such liability to be spelt out in the Conditions of Contract because the parties cannot, by their Contract, alter or avoid their respective statutory responsibilities or exclude the Health and Safety Executive's powers of intervention.

If and to the extent that special safety provisions are required in connection with a specific project, such provisions can properly and appropriately be included in the Employer's Requirements (which, like the Conditions of Contract, is itself a Contract document) or in the Health and Safety Plan.

Clauses 20 to 25: General

Care of the Works, Indemnities and Insurances

Attention is drawn to the fact that the liability clauses - Clause 20 in respect of the Works and Clause 22 in respect of third-party responsibilities - place full responsibility on the Contractor (save, in respect of the Works, for the Excepted Risks listed in Clause 20(2) and, in respect of third-party liabilities, for the exceptions listed in Clause 22(2)).

Clauses 21 and 23 provide safeguards for the Employer by ensuring, through the insurances that the Contractor is required to provide (up to the limits of indemnity stated in the policies), that the Contractor will be able to meet any obligations that arise and for which he is responsible under Clauses 20 and 22.

The Contractor may have a limit of indemnity on his third-party policy that is in excess of the limit stated in the Contract. This would provide the Contractor (and therefore the Employer) with additional protection because the Contractor's liability is not limited to the minimum amount stated in the Contract. However, the Contractor in all cases remains liable for the total amount should any claim be settled at an amount in excess of the minimum amount of insurance cover required under the Contract (Clauses 21(1) and 23(1)).

The Contractor's responsibility to carry employer's liability insurance is a statutory obligation in the UK and thus needs no provision in the Conditions of Contract.

No reference is made to professional indemnity insurance for any design work required under the Contract. Design of Temporary Works that is incidental to work being carried out under the Contract is normally covered under the policy provided for the insurance of the Works (Clause 21). Design liability for any part of the Permanent Works, if it is to be covered by insurance, would have to be covered under a separate special professional indemnity insurance policy agreed by the Parties. See also Section 6.5 of these notes.

Clause 20: Care of the Works

Under Clause 20(1)(a) the Contractor is made fully responsible for the care of the Works and for all materials, plant and equipment for incorporation therein from the Works Commencement Date (defined in Clause 41(1)), normally until substantial completion.

Under Clause 20(1)(a) the Employer takes over full responsibility for completed work from the date of issue of a Certificate of Substantial Completion. It is therefore essential for the Employer's Representative to give the Employer advance notice of his intention to certify substantial completion so that any requisite insurance can be arranged.

Should issue of the Certificate be delayed for any reason then the Contractor remains responsible for care of the Works but may choose to seek a remedy in damages if the delay was unreasonable.

Clause 21(1): Insurance of Permanent and Temporary Works etc.

The Contractor's insurance policy covering the Works must be for the full replacement cost plus an additional 10% to cover additional costs that may arise incidental to the rectification of any loss or damage, but there is no requirement under the Contract for the Contractor to insure his own plant and equipment.

The Contractor may, for his own protection, have some insurance cover for his own plant and equipment against loss through accident, fire, theft and the like.

Clause 25(1): Evidence and terms of insurance

It is in the Employer's interests that insurance policies taken out on his behalf are verified as meeting the requirements of the Contract and that there is evidence of payment of premiums.

Clause 25(2): Excesses

The excesses carried by a Contractor on his insurance policies must be disclosed at the time of tender by his entering the relevant amounts in Part 2 of the Appendix to the Form of Tender. The excesses (or uninsured losses) that the Contractor carries under his insurance policies are the Contractor's own responsibility. Although it should be verified that the excesses quoted are not too high for a particular project, it would not be appropriate for them to be queried unless they appear to be unduly large in relation to the known resources of the Contractor.

Clause 26: Giving of notices and payment of fees

Unless indicated otherwise in the Contract, all notices will be given by the Contractor and any fees required will be paid by the Contractor with the cost being reimbursed under Clause 26(2).

It is not intended that the Contractor should be responsible for obtaining any planning permission in respect of designs provided by others; e.g. as provided in the Employer's requirements. He is responsible for any further permissions needed as a result of his final designs.

If any planning permission as required under sub-clause 26(3)(c) has not been obtained prior to commencement of work on Site or in due time the warranty given ensures that the Contractor will be granted an extension of time for completion of the Works and be reimbursed for any additional costs incurred should there be any delay in the granting of the planning permission required.

Clause 27: New Roads and Street Works Act, 1991

The Contractor must be given full details in the Employer's Requirements of all conditions or limitations imposed by the highway authority or by any other responsible authority when a licence is granted or consent is given for the carrying out of street works covered by the Act. Other authorities may include but are not limited to traffic, bridge, sewer and transport authorities.

Statutory obligations arising under the New Roads and Street Works Act 1991 may extend beyond the date of issue of the Defects Correction Certificate.

Clause 28: Patent Rights

The Contractor's Submission with his tender should identify any patented items of materials or plant for incorporation in the Works and that could involve the Employer in an ongoing obligation.

Clause 36: Quality of materials and workmanship and tests

Quality Assurance procedures are referred to in Clause 8(3). Clause 36(3) requires that the Contractor submit his proposals to the Employer's Representative for approval. The Contractor remains responsible for carrying out the checks and tests.

Clause 41: Works Commencement Date

The Works Commencement Date is defined as one of three alternatives and is the starting date for the time allowed for the completion of the Contract, irrespective of when work actually commences. Since the Works include design it may be some months before construction work commences on Site although access may be needed for survey or further investigation.

Clause 42: Possession of Site and Access

Any restrictions on access will have been set out in the Employer's Requirements. The Contractor's Submission may also include access requirements to be met by the Employer.

Clause 43: Time for completion

Either the Employer or the Contractor is required to state in the Appendix to the Form of Tender the completion time for the whole of the Works or completion times for each of the separately identified Sections comprising the whole of the Works.

Clause 44: Extension of time for completion

Under Clause 44(2) the Employer's Representative must make an assessment of the delay suffered by the Contractor upon receipt of any particulars of delay under Clause 44(1). In addition, in the absence of any notification, the Employer's Representative may make a similar assessment if he considers that the Contractor has in fact suffered a delay. In either event the Contractor must be notified in writing of the result of the assessment.

This requirement is included to ensure the timely recording by the Employer's Representative of any delay suffered or claimed to have been suffered by the Contractor, but there is no need for any decision to be made at that time as to whether any extension of time for completion of the Works should be granted.

Under Clause 44(3) the Employer's Representative has a duty to grant an interim extension of time for completion as soon as he considers that such an extension is due. This is to ensure that the Contractor does not incur additional cost by accelerating progress unnecessarily due to doubt as to whether or not an extension of time will be granted. If a claim for an extension of time is rejected the Contractor must be notified without delay.

Clause 46: Rate of Progress

If the Contractor's progress is too slow to achieve completion by the due date then he is required to expedite his rate of progress and will not be entitled to additional payment. If, however, the Employer requires that the Works be completed earlier than the time for completion set by or arising under the Contract, acceleration measures can be agreed between the parties.

Clause 47: Liquidated damages for delay in Substantial Completion

Clause 47 covers in detail the different situations that can exist either in dealing with the whole of the Works (Clause 47(1)) or where the Works is divided into Sections (Clause 47(2)). There is also provision for a reduction in the damages payable should any part of the Works or part of a Section of the Works be certified as complete earlier than the rest.

It must be noted that the liquidated damages for delay quoted in the Appendix to the Form of Tender are alternatives and that if completion dates are quoted or required for Sections then liquidated damages should not be quoted for failure to complete the whole of the Works within the overall time for completion.

Provision is made in Clause 47(4)(a) for a limit to be placed on liquidated damages if this is thought to be appropriate.

Clause 47(4)(b) provides that if no sum or sums for liquidated damages are quoted in the Appendix to the Form of Tender then no damages, whether liquidated or unliquidated, will be payable.

Clause 47(6) sets out the procedure to be followed if, after liquidated damages have become payable, the Employer's Representative orders an alteration under Clause 51 or accepts that a Clause 12 situation or any other situation outside the control of the Contractor exists that in either event will involve further delay to that part of the Works.

In such a case the Employer's entitlement to liquidated damages is suspended until the further delay comes to an end. Both the commencement and termination of such further delay must be notified in writing by the Employer's Representative.

Clause 48: Substantial Completion

The final tests referred to under Clause 48(1) may include any performance or similar tests set out in the Employer's Requirements.

Prior to issue of a Certificate of Substantial Completion, any necessary operation and maintenance manuals must be provided in sufficient detail to enable the Employer to take over and run the Works.(Clause 48(3)). Any particular requirements for such manuals should be set out in the Employer's Requirements.

Clause 49: Work outstanding

It should be noted that these Conditions of Contract are not suitable for landscaping work or tree planting etc. for which an extended period of care and maintenance is usually required. Such work may be best carried out under a separate arrangement.

The Defects Correction Period will be normally one year. However, where the Contract includes a substantial element of installed plant, different requirements may be appropriate so as to ensure an adequate period of running and testing for any plant. Such specific requirements should be stated in the Employer's Requirements.

Clauses 51:Ordered alterations

Clause 51(1) specifically permits alteration of the Employer's Requirements (including during the Defects Correction Period) but, due to the nature of this Contract, such alterations should be kept to a minimum.

Alterations under this clause are to the Employer's Requirements and not to the Works as designed by the Contractor. When alterations are instructed, the Contractor remains responsible for designing and constructing the Works to meet the altered requirements.

Since the Contractor is responsible for design of the Works he is in a good position to identify potential improvements or cost saving measures. Clause 51(3) thus allows for proposals to be made by the Contractor for alterations to the Employer's Requirements. If accepted these would be ordered under Clause 51(1).

Clause 52: Valuation of ordered alterations

Since the nature of this contract is that the price is based on a Sum as tendered, there are no billed or other individual priced items to assist in valuing alterations. Where alterations can be foreseen, the Contractor is asked to submit his quotation for the altered work and his estimate of any delay, thus allowing matters to be agreed before the work starts.

Where the alteration arises from a Contractor's proposal, provision is made in Clause 52(2) for sharing any changes in value or time fairly between the parties.

In situations where an alteration has already been ordered the Contractor is still required to submit his quotation for the change for acceptance or negotiation with the Employer's Representative. If no agreement is reached the Employer's Representative notifies the Contractor of the figure he considers reasonable for payment purposes. If the Contractor remains dissatisfied he may pursue this under Clause 66.

Clause 52(4) allows for some alterations, usually of a minor nature, being valued on a dayworks basis.

Clause 53: Additional Payments

The procedure for claiming additional payments under the Contract is set out in detail in Clause 53. These procedures and the timing involved are intended to assist resolution of such matters at an early stage. The Contractor's right to payment is subject to his providing details of his claim to the satisfaction of the Employer's Representative.

Clause 54: Non-removal of Materials and Contractor's Equipment

It should be noted that there is no provision for goods or materials owned by the Contractor becoming, when on Site, the property of the Employer. Provision is made, however, for vesting in the Employer goods and materials not on Site, where it has been agreed that payment will be made on account of such items.

Clause 58:Use of PS and PC items

In general the Employer's Requirements should normally cover everything that he wishes to have provided, so that Provisional Sums and Prime Cost items should not be necessary. Should any Provisional Sum or Prime Cost items be included in the Contract, they can only be used with the consent of the Employer's Representative. Valuation of such work will generally be in accordance with Clause 52.

Clause 60: General

Payment provisions

The payment provisions are now intended to take account of the requirements for the Housing Grants, Construction and Regeneration Act 1996 (HGCR Act) with respect to timing and the provision of information.

The Contract must include a method of allowing interim payments. This may be by a payment schedule or other agreed means. The Contractor will then submit his interim statements to the Employer's Representative in accordance with the agreed method and including the matters listed in Clause 60(1) (a) to (d).

A payment schedule could be included in the Employer's Requirements or as part of the Contractor's Submission. In the absence of an agreed schedule the Employer's Representative will assess amounts due on the basis of the Contractor's monthly statement.

Clause 60(2): Interim payment

This Clause maintains the requirement that the final date for payment by the Employer against an Employer's Representative's Certificate is 28 days after the date of delivery of the Contractor's monthly statement to the Employer's Representative.

There is however an additional requirement whereby the Employer's Representative has to issue and deliver his Certificate to the Employer (with a copy to the Contractor) within 25 days of the date of delivery of the Contractor's monthly statement to the Employer's Representative. See also notes to Clause 60(9)

Payment to the Contractor becomes due on certification by the Employer's Representative with the final date for payment being within 28 days of delivery of the Contractor's monthly statement to the Employer's Representative.

These timings are intended to provide compliance with the HGCR Act.

Clause 60(5): Retention

The rate of retention to be deducted in accordance with Clause 60(5) is to be stated in the Appendix to the Form of Tender, Part 1 and it is recommended that this should not exceed 5% . The limit of retention has also to be stated in the Appendix to the Form of Tender, Part 1 and it is recommended that this should not exceed 3% of the sum stated in the Form of Tender.

Clause 60(6): Payment of retention

Half of the retention held is to be released upon issue of a Certificate of Substantial Completion for any part or Section in relation to the value of that part or Section. No payment certificate is required to initiate this release but subsequent certificates under Clause 60(2) should reflect the release.

The remainder of any retention is released at the expiry of the last of the Defects Correction Periods. Should any work be outstanding at this time, the Employer may withhold sums under the provisions of Clause 60(10).

Clause 60(7): Interest on overdue payments

Clause 60(7) states that interest is due to be paid for each day a payment is overdue and that the interest is to be compounded monthly.

As the Employer's Representative is required to certify on the basis of the statement submitted by the Contractor under Clause 60(1), the Contractor when submitting each monthly statement may include the amount of all interest that he claims as due under Clause 60(7) up to that statement date in his valuation under Clause 60(1)(d).

The bank whose base lending rate is to be used for the calculation of interest must be specified in the Appendix to the Form of Tender.

Clause 60(7) makes provision for the payment of interest on sums payable to the Contractor after certificates have been corrected by an arbitrator.

Clause 60(9): Certificates and payment notices

This Clause provides that the Employer's Representative's certificate shall also serve as the Employer's notification to the Contractor of the amount to be paid and the basis on which it was calculated. As the Employer's Representative's certificate has to be issued at least 3 days before the final date for payment the Contractor has at least 3 days notice of any reductions made by the Employer's Representative from the amount claimed by the Contractor in his statement.

If information on the basis of calculation is not provided with the certificate, the Contractor is entitled to ask for full details as to how the amount certified by the Employer's Representative has been arrived at. The Contractor may well need such information so as to be able properly to make payments at the time they are due to any sub-contractors employed on the Contract. Details of any valuation of alterations for new or changed items could also be necessary so that the Contractor can decide whether to indicate his dissatisfaction, with the valuation and the certificate issued, under the provisions of Clause 66(2).

Clause 60(10): Notice of intention to withhold payment

The HGCR Act 1996 requires that any Employer intending to withhold payment from an amount certified must so inform the Contractor. This Clause thus requires that the Contractor be notified by the Employer at least one day before a payment is due if the payment being made will differ in any respect from the amount certified by the Employer's Representative. Details of how any amount being withheld has been calculated and the grounds for so doing must be stated.

Clause 61: Defects Correction Certificate

At the end of the last period for correction of defects and when all the Contractor's obligations have been completed, the Employer's Representative is to issue a Defects Correction Certificate to confirm this. Clause 61(3) requires that finally approved versions of any record drawings or manuals must be submitted prior to issue of the certificate.

Clause 62: Urgent repairs

This Clause allows the Employer's Representative to request that the Contractor carry out any urgent repairs. If the Contractor is unable or unwilling to carry out these repairs, the Clause authorises the Employer to do so and to charge the Contractor for any part of the repairs that were the Contractor's liability.

Clause 64(4): Default of Employer

This Clause makes provision if the Employer is in default and sets out the events upon which the Contractor may rely to terminate his employment under the Contract after giving due notice to the Employer. This Clause balances the provisions of Clause 65 that covers any default of the Contractor.

Clause 65: Default of Contractor

Should the Contractor be in default in any of the ways detailed in Clause 65(1) then the Employer is entitled to enter the Site to expel the Contractor and then complete the Works by other means. In that event the Contractor will not be entitled to further payment until the Employer's Representative certifies under Clause 65(5)(b) the difference between what would have been payable to the Contractor and the actual cost to the Employer of achieving completion. If the difference is an amount due to the Employer then he may recover this from the Contractor or his bondsman.

See Appendix 1 to these notes regarding payments from the Surety under the Performance Bond.

Clause 66: General

Avoidance and settlement of disputes

The Clause is drafted to comply with the adjudication provisions of the Housing Grants, Construction and Regeneration Act 1996. The intention is that the adjudication procedure provided for by the Act's associated "Scheme for Construction Contracts" should not apply.

The HGCR Act introduced procedures for adjudication in construction contracts. These procedures apply to disputes that have arisen between the Employer and the Contractor. It is thus necessary to identify that a dispute has arisen before such a matter can be taken on to adjudication or indeed to any of the means of resolving disputes under this clause. The procedure under Clause 66(2) is part of the mechanism agreed by the parties for identifying disputes. Since much time and money can easily be wasted arguing about issues that have not been clearly defined, the procedure of Clause 66(2) is also seen as a means of clarifying the matter at issue. If either the Employer or the Contractor decide that they can not accept the Employer's Representative's decision under Clause 66(2), or a decision is not given within the time allowed, then they can consider that a dispute exists and issue a Notice of Dispute under Clause 66(3).

Clause 66(1): Avoidance of disputes

The disputes procedure now includes options for both conciliation and adjudication before there is a reference to arbitration. The intention behind Clause 66 is to overcome where possible the causes of disputes and, in those cases where disputes are still likely to arise, to facilitate their clear definition and early resolution preferably by agreement. It is possible there could be consequential savings in both time and cost.

Clause 66(2): Matters of dissatisfaction

If at any time the Employer or the Contractor is dissatisfied with any decision, opinion, instruction, direction, certificate or valuation of the Employer's Representative, or of one of his assistants, or with any other matter in connection with or arising out of the Contract, or the carrying out of the Works, the matter may be referred to and decided by the Employer's Representative.

The Employer's Representative must give his decision on all matters included in the written submission of the referring party within one month after receiving that written submission. This decision of the Employer's Representative is an essential preliminary to establish if a dispute exists and, if so, exactly what is in dispute, before the disputes procedure set out in the rest of Clause 66 can be operated. It is expected that, in practice, most matters will be resolved at this stage.

Clause 66(3): Disputes This Clause sets out what can give rise to or constitute a dispute or difference leading to a Notice of Dispute. The Contractor and Employer agree by this clause that they will not consider themselves to be in dispute until a Notice of Dispute has been issued under this clause and that the Notice defines what is in dispute.

Clauses 66(5) to 66(9): General ***Settlement of disputes***

Once a valid Notice of Dispute has been issued under Clause 66(3), there are options for conciliation or adjudication before there is a reference to arbitration. It is anticipated that most disputes will be dealt with as soon as possible after they arise. With the potential for the prompt settlement of many minor disputes or differences that can arise on a Contract this should lead to considerable savings in time and cost.

Clause 66(5): Conciliation A dispute can be referred to conciliation at any time after the Notice of Dispute has been issued unless a Notice to Refer that same dispute to arbitration has already been issued. Thus, a conciliation under this Clause can proceed in parallel with any reference to adjudication or with an arbitration commenced after the reference to conciliation.

If the dispute is not settled by agreement during the conciliation, the conciliator will issue a recommendation that will become final and binding unless the dispute is referred to adjudication or direct to arbitration within one month of receipt of the recommendation by the dissenting party.

Clause 66(6): Adjudication The Contract incorporates by reference the ICE Adjudication Procedure (1997) (or any amendment or modification in force at the time of the Notice of Adjudication).

Following a Notice of Adjudication, 7 days are allowed for appointing the adjudicator and referring the dispute to him, whereupon he is given 28 days (or such other period as may be permitted) in which to reach his decision.

Once a decision has been reached and notified to the parties it becomes binding unless and until the dispute (not the decision) is referred to arbitration.

A failure to give effect to a decision is excluded from the arbitration agreement leaving the parties free to seek enforcement through the courts.

If a dispute on which a decision has been given is not referred to arbitration by issuing a Notice to Refer under Clause 66(9) within three months of the giving of the decision, that decision becomes final as well as binding and cannot thereafter be challenged.

Clause 67(2): Application to Scotland Following publication of "The Scottish Arbitration Code 1999", the Sponsoring Bodies have agreed that that Code should be adopted for ICE arbitrations in Scotland. Accordingly, the ICE's "Arbitration Procedure (Scotland) (1983) has now been withdrawn and, in its place, the 1999 Code will apply as modified by the ICE's new "Appendix (2001) to The Scottish Arbitration Code 1999" that is published separately. In particular, Sub-clause 67(2)(f) deals with the method of enforcement of an arbitrator's award or an adjudicator's decision in Scotland and the Appendix Item 19 is not required in Scotland.

Clause 67(3): Application to Northern Ireland

The provisions of the Housing Grants, Construction and Regeneration Act 1996 (HGCR Act 1996) have been covered by amendments to various Clauses in these Conditions of Contract but when the Construction Contracts (Northern Ireland) Order 1997 (CC(NI) Order 1997) was published it had a different clause numbering system from that used in the HGCR Act 1996.

For ready reference the Sections of the HGCR Act 1996 are cross referenced below to the applicable Paragraphs of the CC(NI) Order 1997. It should however be noted that there are differences in the text between the Act and the Order. The HGCR Act 1996 came into force on 1 May 1998. The CC(NI) Order 1997 came into force on 1 June 1999.

HGCR Act 1996 Section	Construction Contracts (NI) Order Paragraph
104	3
105	4
106	5
107	6
108	7
109	8
110	9
111	10
112	11
113	12
114	13 and 16
115	14
116	14(c)
117	15

Clause 69(1)(b): Labour and landfill tax fluctuation

The new Clause 69 applies also to the Finance Act 1996 (Sections 39-71 and Schedule 5) and the Landfill Tax Regulations 1996. It is made clear that only those rates of Landfill Tax chargeable under the above Act and Regulations in force at the date for return of tenders shall be deemed to have been included in the tender rates, whether or not future increases or decreases are then foreseeable. Increases or decreases after the date for return of tenders are covered in Clause 69(2).

Clause 71: CDM Regulations 1994

The Construction (Design and Management) Regulations 1994 apply to all construction contracts, other than:

Site surveys.

Exploration and abstraction of mineral resources and activities preparatory thereto carried out at the site of such exploration or abstraction.

Projects for which the construction phase will not last more than 30 days and which do not involve more than 500 man days of construction.

Projects for which the local authority is the enforcing authority under the Health and Safety (Enforcing Authority) Regulations 1989.

Construction (but not demolition) where not more than 5 persons are employed at any one time.

Construction for a private client in respect of his own dwelling.

Anybody entering into a contract where the CDM Regulations apply should study the Regulations in detail. The list of exclusions above is not exhaustive.

Under these Regulations the Employer is required to appoint, as soon as practicable, a PLANNING SUPERVISOR responsible, among other things, for the preparation and monitoring of a HEALTH AND SAFETY PLAN for the contract, or for the project of which the contract is a part.

The name of the Planning Supervisor is included in Part 1 of the Appendix to the Form of Tender, and the Health and Safety Plan for the Works and the Health and Safety Files in respect of each structure included in the Works (to the extent that they have been prepared) will be available for inspection by tenderers.

Under these Regulations the Employer is also required to appoint for the contract as soon as practicable a PRINCIPAL CONTRACTOR with particular responsibilities regarding safety whose name (if one has been appointed by that time) is also included in Part 1 of the Appendix to the Form of Tender.

The Contractor could also be asked to act as Principal Contractor and may also be asked to act as Planning Supervisor.

On completion of the Contractor's obligations under the Contract, the Safety File becomes the property of and must be handed over to the Employer.

Where these responsibilities are not covered by separate contracts they are likely to be covered in the Specification with the cost of compliance covered by the Contract Price.

Clause 72: General

Special Conditions

Special conditions may be added if essential to cover specific requirements.

Contract Price Fluctuation Clauses

Contract Price Fluctuation Clauses are printed in the Design and Construct Conditions of Contract document and are also supplied in loose-leaf form. When one (or more) is to be incorporated in a Contract it should be included as a special Condition.

If a Contract Price Fluctuation Clause is to be included in a Contract attention is drawn to Appendix 2 of these Guidance Notes. This Appendix sets out the views of the CCSJC on what constitutes good practice when a Contract Price Fluctuation Clause is used.

APPENDIX 1

PERFORMANCE BOND

General The Bond may be called only if the Contractor is expelled from the Site as a result of insolvency or is removed from the Site under any other of the provisions of Clause 65. If the Contractor has not been expelled from the Works, there is sufficient redress in the Contract (e.g. retention, set-off, damages etc.) to deal with other matters that might constitute a default. The Bond relates to Clause 65 for matters of expulsion and certification subsequent to that expulsion, and to the Contractor's failure to pay the Excess Sum (or any part of the Excess Sum remaining unpaid).

Surety Bond The reference to Clause 66 enables the Surety, if he so wishes, to utilise the same adjudication provisions which are available to the Contractor in order to challenge the Employer's Representative's Certificate.

Although the Surety has the right to receive copies of any notice served on the Contractor under Clause 65(1) or any certificate served under Clause 65(4) or 65(5), and to request information or to inspect the Site, this should not be considered to be a condition precedent to payment. The only condition precedent to payment by the Surety is the service on the Surety of the **certified copy** Clause 65(5) certificate. This certificate must be certified by the Employer's Representative as a true copy and it may be an interim certificate or a final one.

By accepting this bond, the Surety undertakes, within 14 days of receiving the certified copy of a Clause 65(5) certificate, either to pay or to refer the matter to adjudication. If the Surety wishes to contest the certificate or if the Employer refuses for example to allow the Surety's representative to inspect the Site or provide information which the Surety requests, he must seek the decision of the adjudicator in the first instance under the provisions of Clauses 66(6) to 66(8) of the Contract. The adjudicator's decision is binding unless and until the dispute is finally determined by legal proceedings, by arbitration or by agreement.

The provision in clause 1(6) of the Bond in relation to a previous adjudication between the Employer and the Contractor prevents a second adjudication from taking place on the same matter. However a safeguard is incorporated to the effect that both parties must have made submissions to the adjudicator before his decision was given. This is necessary in the event that the Contractor is unable to respond to an adjudication.

The time periods have been set so that payment is not delayed unnecessarily in the event of there being no dispute. On the other hand, in the event of a dispute, the Surety has sufficient time to protect his interests. Clause 1(7) of the Bond sets the time for payment at 7 days from the decision of the adjudicator in the event that there is no other direction.

APPENDIX 2

CONTRACT PRICE FLUCTUATION

Foreword Three Contract Price Fluctuation Clauses are printed in the Design and Construct Conditions of Contract for inclusion in a Contract as special conditions when required. Loose leaf copies are also available. The Index Figures used are compiled by the Department of the Environment Transport and the Regions and published by The Stationary Office in the Monthly Bulletin of Indices "1990 Series Civil Engineering Formula Indices" and "1990 Series Structural Steelwork Formula Indices".

The CEW Clause is provided for use in connection with Works of Civil Engineering Construction, the SS Clause is provided for use on Contracts that are wholly or predominantly in respect of Structural Steelwork and the CEW/SS Clause has to be incorporated with both the CEW Clause and the SS Clause when provision is to be made for the adjustment of price fluctuation on Contracts involving both Civil Engineering Work and Structural Steelwork.

In the Guidance Notes that follow, reference to the Contract Price Fluctuations Clause (CPF Clause) is to be taken to refer to the CEW Clause or to the SS Clause or to the CEW, the SS and the CEW/SS Clauses.

1 Introduction The CPF Clause offers valuable savings in administrative effort compared with earlier variation of price clauses. These savings are achieved because the complexities arising from the effects of price variations on a contract are reduced to one simple formula. To achieve this degree of simplification many approximations and generalizations must be made: for example, the Employer's assessment of the proportions in which different types of work contribute to the Works as a whole and the assumption that these different types of work will be performed at a steady rate throughout the period of the Contract.

It is necessary for both the Employer and the Contractor to accept that degree of approximation that is inherent in the CPF Clause. It is equally important that particular aspects of the CPF Clause should not be viewed in isolation as being likely to produce advantages or disadvantages to either the Employer or the Contractor.

Where contracts are awarded by competitive tendering any advantages or disadvantages that tenderers foresee arising from the approximations inherent in the CPF Clause will be reflected in their tenders and are unlikely to give rise to inequitable payments.

It is desirable that a uniform method of operating the CPF Clause should be established to enable tenderers to base their tenders on uniform precepts. The CCSJC is unanimous that in most contracts the methods of operating the CPF Clause set out in sections 2-4 of this Appendix are good practice and in accordance with business sense.

If, in an exceptional case, the CPF Clause is to be operated in a different manner, the proposed alternative should be described in the tender documents to enable tenderers to take its effects fully into account in the Contract.

2 Retention

The CCSJC takes the view that the contractual entitlement to any payment is dependent upon Clause 60(2) and that the CPF Clause only provides for a method of calculating the fluctuation payments due as part of that payment. Payments due under the CPF Clause will be shown by the Contractor in his monthly statement in accordance with Clause 60(1)(d). Appropriate payments will be certified by the Employer's Representative in accordance with Clause 60(2) and will thus be subject to retention in accordance with Clause 60(5).

3 Dayworks and Nominated Sub-contractors

The CCSJC considers that the words "... based on actual cost or current prices ..." in CEW Clause (2)(d)(i) apply to both Dayworks and Nominated Sub-contractors. Thus, amounts for Dayworks, or parts thereof, and Nominated Sub-contractors, where paid at actual cost or current prices, should be excluded from the Effective Value while those amounts based on rates and prices quoted at the time of tender should be included in the Effective Value.

4 Goods and materials on Site or vested in the Employer

The amounts included in an interim certificate pursuant to Clause 60(2) in respect of goods or materials on Site or vested in the Employer are normally a percentage (in no case exceeding the percentage stated in the Appendix to the Form of Tender) of the invoiced prices of the goods or materials.

When calculating the Effective Value, the above amounts, because they are based on the invoiced prices and thus items based on actual cost, should be deducted under CEW Clause (2)(d)(i) from the amount which, in the opinion of the Engineer, would otherwise be due to the Contractor under CEW Clause 2(d)(i), i.e. they do not at that stage form part of the Effective Value at that time.

In the month in which any of these goods or materials are incorporated into the Permanent Works the value of those goods or materials at tender prices (being part of a schedule payment or some other means of payment) will form part of the estimated contract value of the Permanent Works certified under Clause 60(2) and thus become part of the Effective Value at that time. The Contractor will thus be paid an amount in respect of the tender value of the goods or materials under Clause 60(2) plus a fluctuation payment under CEW Clause (2)(d)(i) increasing or decreasing the tender value of the goods or materials in line with the increase or decrease in the value of the Works as a whole as reflected through the Price Fluctuation Factor. The percentage of the invoiced price previously paid will effectively be deducted as a previous payment on account through the operation of Clause 60(2).

5 Comment

Attention has been directed to the case where, during a period of high inflation, goods and materials remain on Site for a considerable time, during which the Price Fluctuation Factor increases significantly. Although, as stated in section 1, particular aspects of the CPF Clause should not be viewed in isolation the CCSJC considers that this matter requires special mention. It has been suggested that in the circumstances described the Contractor would be paid an unjustifiably higher price for goods and materials than he need have been paid had the goods and materials been subject to the Price Fluctuation Factor of the month in which the goods or materials were included in an interim certificate.

However, when goods or materials are brought on to Site expediently it contributes to timely completion and for this added safeguard the Employer has paid only a percentage of the invoiced price of the goods or materials in advance of their incorporation into the Works. The longer the goods or materials remain on Site the longer is the Contractor obliged to finance that percentage of the goods or materials which has not been paid for by the Employer. The Contractor is also responsible for replacing any goods or materials stolen or spoilt while on Site. Viewed in the context of the CPF Clause the CCSJC considers that neither party receives inequitable treatment.

ICE Conditions of Contract

Seventh Edition

Measurement Version

Guidance Notes

The Institution of Civil Engineers
The Association of Consulting Engineers
The Civil Engineering Contractors Association

Published by Thomas Telford Publishing, Thomas Telford Limited, 1 Heron Quay, London E14 4JD (Telephone +44 (0)207 987 6999 Fax +44 (0)207 538 5746 E-mail http://www.t-telford.co.uk) on behalf of:

The Chief Executive and Secretary
The Institution of Civil Engineers
One Great George Street
London SW1P 3AA

The Chief Executive
The Association of Consulting Engineers
Alliance House
12 Caxton Street
London SW1H 0QL

The Director
Civil Engineering Contractors Association
Construction House
56-64 Leonard Street
London EC2A 4JX

Guidance Notes to the ICE Conditions of Contract 7th Edition September 1999

9 8 7 6 5 4 3 2

ISBN 0 7277 2842 3

A catalogue record for this book is available from the British Library

Printed and bound in Great Britain by Selwood Printing, West Sussex

CONTENTS

Associated publications referred to in the Guidance Notes

ICE Conditions of Contract – Measurement Version 7th Edition

ICE Arbitration Procedure 1997
ICE Arbitration Procedure (Scotland) (1983)
ICE Conciliation Procedure 1999
ICE Adjudication Procedure 1997

Other ICE Conditions of Contract publications

ICE Design and Construct Conditions of Contract
Guidance Notes to the ICE Design and Construct Conditions of Contract
ICE Conditions of Contract Minor Works 2nd Edition (including Guidance Notes)

INTRODUCTION

Standard agreed forms of conditions of contract (referred to as the ICE Conditions of Contract) for use in connection with works of civil engineering construction have been published since December 1945.

After the publication in June 1973 of the Fifth Edition of the ICE Conditions of Contract a permanent joint committee - the Conditions of Contract Standing Joint Committee (CCSJC) - was set up to keep the use of standard forms of contract under review, to consider any suggestions for amendment and also to undertake the drafting of new standard forms.

The composition of the CCSJC is as follows:

An independent chairman appointed by the ICE.
Three members appointed by the ACE to represent Consultants.
Three members appointed by the CECA to represent Contractors.
Three members appointed by the ICE on the advice of client bodies to represent Clients.
One co-opted member (a Fellow or Member of the ICE with an academic or legal background) appointed by the ICE.

The terms of reference of the CCSJC exclude the provision of any legal interpretation.

These Guidance Notes have been prepared by the CCSJC specifically to assist users of the ICE Conditions of Contract in the preparation of contract documents and the carrying out of the contract Works. They do not purport to provide legal interpretation but do represent the unanimous view of the CCSJC on what constitutes good practice in the conduct of civil engineering projects.

Like all previous editions of the ICE Conditions of Contract, the Seventh Edition is based on the traditional pattern of Engineer-designed, Contractor-built Works with valuation by admeasurement. The traditional role of the Engineer in advising his client, designing the Works, supervising construction, certifying payment and deciding what are now called matters of dissatisfaction is fully maintained.

The Contract is drafted bearing in mind the benefits of team working and current procurement initiatives. If its procedures are followed the parties to the Contract will be provided with an "early warning" of circumstances that may give rise to additional costs or delay in a stage or completion of the Contract

To enable the minimisation of additional cost and/or delay as well as potential for dispute all possible steps should be taken to avoid changes to pre-planned Works information. Design work should be completed before tender information is sent out and then not be subject to change. Changes in requirements after contracts have been let will normally lead to substantial increases in cost and undermine the principle of mutual co-operation.

The Contract is also drafted with the intention of minimising the incidence of potential disputes and resolving those that may arise in a speedy and non-confrontational manner. The Contract provides the choice of conciliation, adjudication and arbitration to facilitate the resolution of disputes.

The Housing Grants, Construction and Regeneration Act 1996 came into force on 1 May 1998 for construction contracts as defined in the Act. The Contract is drafted to comply with this legislation with the intention that the Statutory "Scheme for Construction Contracts" should not apply (see in particular Clauses 60 and 66). Clause 67 deals with the application of the legislation to Scotland and the Construction Contracts (Northern Ireland) Order in Northern Ireland.

Compared with previous editions, several substantial changes and a number of minor ones have been made and a certain amount of editing has been carried out to improve the clarity of the document.

It should be noted that these Guidance Notes should not and do not form part of the Contract.

CONTRACT DOCUMENTS

When Clauses 1 to 72 of the ICE Conditions of Contract are incorporated in a Contract it is recommended that they are incorporated unaltered because they are closely interrelated and any changes made in some may have unforeseen effects on others.

If it is necessary on a particular Contract to make any special contractual arrangements these should be effected by the addition of special conditions in accordance with Clause 72 of the ICE Conditions of Contract. Special conditions should be kept to the minimum necessary to cover the special circumstances of the particular project. The practice of including as special conditions matters which are more appropriate to the Specification should be avoided.

This procedure assists comprehension of the documents, particularly at the tender stage, because it obviates the need for tenderers to scrutinize the standard conditions to check for possible amendments or additions.

TENDER DOCUMENTATION AND PROCEDURES

It is recommended that the procedures set out in the publication "Tendering for Civil Engineering Contracts" be followed.

It is important for Employers to be fully aware of the recommendations made in respect of tendering procedures, and for them to avoid putting undue pressure on Engineers to send out invitations to tender before the design and preparation of the Drawings and other contract documents have been completed. Employers should also allow adequate time for the preparation of tenders.

MATTERS FOR THE EMPLOYER TO DECIDE DURING THE PREPARATION OF THE TENDER DOCUMENTS

Under the ICE Conditions of Contract the Employer has to appoint an Engineer (who may be a person, firm or company) to design the Works, prepare the Contract documents and supervise the construction of the Works as well as generally administering the contract. If the Engineer so appointed is not a single named Chartered Engineer the Engineer has a duty, after the award of the Contract, to notify to the Contractor in writing the name of a Chartered Engineer who will act on his behalf and assume the full responsibilities of the Engineer under the Contract.

One of the first appointments an Employer has to make is that of the Planning Supervisor under the Construction (Design and Management) Regulations 1994 (more commonly known as the CDM Regulations) (Clause 71(1)(b)). Under these Regulations the Planning Supervisor is required, in the first instance, to manage the preparation of the Health and Safety Plan. Other duties are referred to in Clause 71 of these Notes, however, as the Notes are not exhaustive, the Regulations should be studied in detail.

During the preparation of the tender documents the Employer, generally in consultation with or with advice from the Engineer, must decide

(a) the duration of the Defects Correction Period (Clause 1(1)(s) and Part I of the Appendix to the Form of Tender)

(b) the number and type of Drawings to be provided to the Contractor (Clause 6(1)(b))

(c) which parts of the Permanent Works, if any, are to be designed by the Contractor (Clause 7(6))

(d) whether or not a formal Contract Agreement will be required from the successful Contractor (Clause 9)

(e) whether a performance bond will be required from the successful Contractor, and if so its amount (Clause 10)

(f) what information he has relevant to the Works and which is to be passed to the Engineer so that details can be included in the documentation prepared for the Contract. (In this connection it would be advisable for the Employer to note carefully the contents of the Guidance Note for Clause 11)

(g) the minimum level of third party insurance to be called for (Clause 23(3))

(h) the commencement date for the Contract (Clause 41(1)) if this is to be pre-set

(i) what restrictions, if any, apply to the availability of any part or parts of the Site and/or the access to the Site (Clause 42(1))

(j) whether provision should be included for Sectional completion (Clause 43)

(k) the time that should be allowed for completion of the Works or (if appropriate) any Section of the Works (Clause 43)*

(l) the level of, and any limit to, any liquidated damages to be levied for any delay in substantial completion of the Works or (if appropriate) each Section of the Works (Clause 47)

(m) whether provision should be made for payment for materials not on Site under the vesting procedure (Clause 54(4))*

(n) what work, if any, should be carried out by Nominated Sub-contractors (Clause 59) and whether they are required to design the same (Clause 58(3))

(o) the percentage of their value to be paid for materials on site or vested in the Employer (Clause 60(2)(b))

(p) the minimum amount of an interim certificate to be paid (Clause 60(3))

(q) the rate and limit of retention money to be held (Clause 60(5))

(r) the bank whose base lending rate is to be used in calculating any interest due on a late payment (Clause 60(7))

(s) whether to appoint a named Adjudicator to act under the provisions of Clause 66(4)

(t) on the arrangements for the Principal Contractor to be appointed under the Construction (Design and Management) Regulations 1994 (Clause 71(1)(b))

(u) the extent to which the Health and Safety plan for the Works required under the Construction (Design and Management) Regulations 1994 (Clause 71(1)(c)) and the Health and Safety Files in respect of each of the structures included in the Works need to be prepared and to be available for issue to or inspection by tenderers

(v) whether to include a Contract Price Fluctuation Clause.

Any requirements where the Employer's approval has to be obtained before the Engineer can act as required under the Contract must be listed in the Appendix to the Form of Tender (see Clause 2(1)(b)).

* If so decided by the Employer these matters may be left for the Contractor to complete when tendering.

NOTES ON SPECIFIC CLAUSES

Clause 1(1)(a): Employer

The Employer has a number of duties to perform under the Contract.

If a Contract Agreement (Clause 9) is required by the Employer he is responsible for its preparation. The Employer also has responsibility for approving the Contractor's performance security (Clause 10(1)). For his own protection he should also check the terms and coverage of the Contractor's insurance policies and verify that premiums have been paid (Clauses 25(1) and (3)).

The Employer is responsible for ensuring that the Contractor will be given possession of and access to the Site to enable him to commence and proceed with the Works. Any restrictions must be set out in the contract documents. The Employer is also responsible for ensuring that payments are made to the Contractor by the final date for payment following certification by the Engineer.

The Employer has the power, in certain circumstances, to terminate the employment of the Contractor.

Clause 2: Duties and authority of Engineer

The Engineer is appointed by the Employer to supervise the Contract and to that end the Engineer has conferred on him by both Parties the powers and duties vested in him under the Contract.

It is in the interests of both the Employer and the Contractor that any queries or problems that arise during construction should be resolved expeditiously. To this end the Contract requires that the Engineer's function shall be exercised by a single named Chartered Engineer even though the appointed Engineer may have been a person, firm or company.

If Clause 2(2)(a) applies and/or Clause 2(2)(b) could apply and the Employer decides that it will be necessary for a person who is not a Chartered Engineer to be appointed to assume the full responsibilities of the Engineer under the Contract this must be drawn to the attention of Contractors tendering when tenders are invited.

Good practice dictates that the Engineer, when carrying out his functions under the ICE Conditions of Contract, should keep the Employer informed of progress, performance, variations and all other relevant matters.

Engineer's independence and impartiality

The Engineer acts as the agent of the Employer to supervise the construction of the Works and to satisfy himself that the Works are being constructed in accordance with the Contract. Of no less importance are those functions which require the Engineer to make decisions regarding any changes to the Works that need to be made in the course of construction or which arise from the allocation of risks. In all aspects of his duties under the Contract both the Engineer and the Engineer's Representative are required by Clause 2(7) to act in accordance with the terms of the Contract, impartially, towards both parties to the Contract, having regard to all the circumstances.

The Engineer has the duty to exercise his professional judgement, honestly and impartially to decide such matters as unforeseen conditions (Clause 12), extensions of time (Clause 44), the valuation of variations to the Works (Clause 52) and to formally decide any issues where either the Employer or Contractor are dissatisfied (Clause 66).

Provision is made in the ICE Conditions of Contract for the Employer to stipulate that his approval must be obtained before the Engineer exercises his authority on specific matters. The particular matters on which the Engineer can only so act after receipt of approval from the Employer must be clearly set out in the Appendix to the Form of Tender (Clause 2(1)(b)) so that the Contractor is aware of the restrictions imposed at the time of tender.

It follows that if the Employer has placed restrictions on the powers of the Engineer before the Contract is made that prevent the Engineer acting without prior approval, the responsibility for acting within the terms of the Contract is assumed by the Employer. Any action or any failure to act on his part which is not within the terms of the Contract may render him liable for damages.

Clause 3(2): Rights of Third Parties

By this Clause the acquisition of rights by third parties is excluded.

Clause 4: Sub-contracting

Clause 4 recognises the practice in the construction industry of the use of sub-contractors and self-employed labour. In general the Contractor should be free to choose his own sub-contractors.

There are two aspects to Clause 4(2). The first allows the Employer, at the tender stage, to advise the Contractor via the Appendix to the Form of Tender of those aspects of the work which the Contractor must not sub-contract to others without the specific authority of the Engineer.

The second aspect requires the Contractor to advise the Engineer of the name and address of a proposed sub-contractor at least 14 days in advance of that sub-contractor's entry on to the Site. The Contractor should take care not to create a binding obligation to his sub-contractor until he is sure the Engineer will not object.

Any objections to a specific sub-contractor must be made within 7 days of receiving the notification from the Contractor. It is essential that the Engineer not only gives his reasons for objection but also that these must be in writing. The reasons must be good as it should be remembered that there is a possibility that the objection could be challenged. Caution must be exercised in the manner of objection to a specific sub-contractor so that legislation on procurement procedures is not infringed. If a specific sub-contractor is objected to without good reason it is possible the Contractor could be involved in additional costs for which he may be entitled to reimbursement under the Contract.

Under Clause 4(4) the Contractor is fully responsible for all sub-contracted work and for the acts, defaults and neglects of any sub-contractor or his employees. The Engineer's acceptance of a sub-contractor in no way relieves the Contractor of this responsibility.

The provisions of Clause 4(5) (in respect of sub-contracting firms) and Clause 16 (in respect of individuals) give the Engineer power to require the removal from Site of sub-contractors or other persons in the circumstances described.

Clauses containing provisions that could relate to design work that may be carried out by a sub-contractor are as follows:

> Clause 4(2) requires notification of any design sub-contractor upon his appointment.
>
> Clause 4(5) covers the removal from the Works or their design of any sub-contractor in the circumstances described.

Clause 6(2) covers the supply of Drawings, Specifications, and other documents.

Clause 6(3) deals with the question of the copyright of Drawings, Specifications and other documents provided by the Contractor.

Clause 7(2) covers the supply of further documents if required by the Engineer.

Clause 7(6)(a) includes provisions for the submission of the Contractor's design information to the Engineer for acceptance.

Clause 7(6)(b) covers the provision of operation and maintenance manuals.

Clause 7(7) deals with the question of responsibility following approval and also makes it clear that the Engineer is responsible for the integration and co-ordination of the Contractor's design with the rest of the Works.

Clause 8(2) deals with the question of design responsibility. (See also note against Clause 8(2)).

Clause 16 contains provisions similar to those in Clause 4(5) and covers the dismissal from the Works of individuals in the circumstances described.

Clauses 58(3) and 59 deal with design by nominated sub-contractors.

The above provisions cover the employment of sub-contractors in all normal circumstances. If an Employer wishes to exercise total control over the selection of a sub-contractor for employment by the Contractor for part of the Works this should be done by the use of the Nominated Sub-contract procedure (Clause 59) together with a Provisional Sum or Prime Cost item. Alternatively, and subject to any applicable procurement legislation, the Contract may simply specify the use of a particular supplier or sub-contractor and the Contractor would allow for this in his tendered rates and prices.

Clause 8(2): Contractor's design responsibility

The Contractor will not generally be responsible for the design or specification of the Permanent Works, but provision is made whereby the Contractor can be required to design part or parts of the Permanent Works. Under the provisions of Clauses 8(2) and 58(3) the Contractor can be responsible for the design or specification of Permanent Works only to the extent that the Contract expressly so provides.

If the Contractor is made responsible for the design of any part or parts of the Permanent Works, the exclusion of liability for design under Clause 20(2)(b) will, to that extent, not apply and responsibility for the design of such part or parts of the Works will be placed on the Contractor under Clause 20(1) (Care of the Works).

Professional indemnity insurance

Some standard Contractor's All Risks insurance policies (provided as required under Clause 21(1) for the insurance of the Works), will exclude all risks associated with the design of any part of the Permanent Works. Other such policies provide cover against loss of or damage to elements which are properly designed if they are lost or damaged as a result of the defective design of another part of the Works. Damage to parts defectively designed will not usually be covered.

If design responsibility is excluded under the Contractor's standard insurance of the Works policy the Contractor must consider whether to accept any liability resulting from a failure of design himself or whether to cover it by insurance or, if he employs a separate design engineer, whether or not he should require that design engineer to carry insurance against any loss arising as a result of a design failure.

In the absence of special arrangements (such as project insurance taken out by the Employer) cover against damage to a defectively designed part of the Works can be provided only by a professional indemnity insurance policy. Under the Contract, liability for a design failure will still lie with the Contractor but if the Employer requires the risk to be covered by insurance this requirement needs to be set out clearly in the instructions to tenderers and should state the minimum cover to be provided by the policy. The tenderers can also be asked to indicate any excess (or deductible amount) on their professional indemnity insurance policy.

Depending on the way any Contractor's design work is carried out (either by the Contractor in-house or by a separate design engineer) different situations may exist. If the Contractor has a current professional indemnity insurance policy it will normally provide cover to the Contractor up to the policy limit whether the design is carried out by the Contractor in-house or is sub-let to a reputable design engineer. Under the provisions of Clause 4(4) the Contractor remains fully responsible for all sub-contracted work.

Particular points to be noted by Engineers and Employers are as follows.

(a) Under the second part of Clause 8(2) it is made clear that the Contractor must exercise all reasonable skill, care and diligence in his design.

(b) Under Clause 4(2) notification has to be given to the Engineer of the name and address of any sub-contractor to whom any design work is entrusted upon appointment.

(c) The requirement that liability for a failure of design be covered by insurance can be written into the Contract if the extent of design by the Contractor warrants this.

(d) Professional indemnity insurance policies are annual policies which have to be renewed every year and provide cover only while they remain in force. Since the cover under such a policy applies only to claims made in that year and not to those arising from negligence in that year which gives rise to a claim in a later year, the Contractor or design sub-contractor need to establish the duration of the insurance required.

Clause 10: Performance Security

A Provisional Lump Sum Item should always be included in the Bills of Quantities for the provision of the Performance Security required under Clause 10(1). For a Guidance Note on the Performance Bond see Appendix 1.

Clause 11: Provision and interpretation of information

Employers seek certainty of expenditure on a project and to assist in this objective it is beneficial to both Employer and Contractor for all information held to be provided at tender stage. The effect of Clause 11(1) is that information obtained by or on behalf of the Employer relevant to the Works shall only be taken into account to the extent that it was made available to the Contractor before the submission of his tender. The interpretation of all such information remains the Contractor's responsibility. If the Employer fails to make available such relevant information at the proper time he cannot later argue that it ought to have been known to the Contractor. The Contractor would, however, always be expected to have obtained for himself any information that was otherwise readily available and if possible to have visited the Site (Clauses 11(2) and 11(3)).

Clause 12: General

Adverse physical conditions or artificial obstructions.

It is impossible for either of the contracting parties to foresee every circumstance that may affect the construction of the Works. It is, however, an essential feature of the ICE Conditions of Contract that the Contractor accepts full responsibility for physical conditions or artificial obstructions that could reasonably have been foreseen by an experienced contractor. It is the Contractor's obligation to construct the Works and he must therefore propose any measures to deal with the conditions or obstructions encountered, for the Engineer's consent.

The Parties rely on the Engineer's assessment of the effect of the circumstances encountered but the consequences of any unforeseeable risks have to be dealt with separately in accordance with the terms of the Contract. Such circumstances may result in extra cost and/or extra time becoming due to the Contractor.

It is clearly in the Employer's own interests to have an adequate Ground Investigation carried out before tenders are invited.

Clause 12: Adverse physical conditions and artificial obstructions

Clause 12(1) requires the Contractor to give written notification to the Engineer as early as is practicable of any physical condition or artificial obstruction encountered which the Contractor considers he could not have reasonably foreseen whether or not there is any intention to claim additional payment or time. Such notification must give details of any anticipated effects, the measures being taken or proposed and their estimated cost and any consequential delay and/or interference with the construction of the Works.

Notification of the intention to claim additional payment or an extension of time is still required and can be made at the same time or as soon as is reasonable thereafter (Clause 12(2)).

Following receipt of written notification, the options open to the Engineer under Clause 12(4) include requiring the Contractor to investigate any alternative measures that may be available for dealing with the situation.

Clause 14: Programme to be furnished

Clause 14 calls for the submission of a programme and general method statements for construction within 21 days after the award of the Contract (Clause 14(1)). Such a programme and method statements are intended as a management tool to indicate to the Engineer when specific areas of the Site and/or additional information and/or Drawings will be required by the Contractor and to enable an assessment to be made of any failure by the Contractor to maintain the required rate of progress on the Contract(Clause 46(1)). Provision is made for the Engineer to accept the Contractor's programme (Clause 14(2)) and accept the method statements (Clause 14(7)), but not for him to approve them.

After the Contractor has submitted his programme and/or method statements the Engineer has 21 days in which to accept them, reject them or request further information.

In respect of the programme the Contractor then has 21 days in which to submit a revised programme or provide the further information requested. When the Contractor submits a revised programme a further 21 day period is allowed for acceptance or rejection. If, after submission by the Contractor of a Clause 14 programme, the Engineer fails to respond within the 21 days allowed the programme is deemed to have been accepted as submitted.

The same specified restrictions do not apply in the case of method statements and methods of construction (Clause 14(6)) since it may well take longer than a pre-determined time to provide revised calculations etc. or verify that they are in accordance with the Contract.

It is always possible to include in the documentation for a Contract a programme or method statement for carrying out the whole or any part of the Works. Similarly the Contractor can be required to submit his programme or method statement with his tender. Such programmes and method statements could become contractual requirements which the Engineer may not change without the agreement of the Parties. Should any outside influence prevent compliance with any such programme or method statement the responsibility for the delay that occurred or for any additional costs arising as a result of the delay could be the responsibility of the Employer.

Clause 19: Safety and security

Liability for safety is imposed by the Health and Safety at Work etc. Act 1974 and other statutes. It would therefore be inappropriate and could be misleading for such liability to be spelt out in the Conditions of Contract because the parties cannot, by their Contract, alter or avoid their respective statutory responsibilities or exclude the Health and Safety Executive's powers of intervention.

If and to the extent that special safety provisions are required in connection with a specific project, such provisions can properly and appropriately be included in the Specification (which, like the Conditions of Contract, is itself a Contract document) or in the Health and Safety Plan.

Clauses 20 to 25: General

Attention is drawn to the fact that the liability clauses - Clause 20 in respect of the Works and Clause 22 in respect of third-party responsibilities - place full responsibility on the Contractor (save, in respect of the Works, for the Excepted Risks listed in Clause 20(2) and, in respect of third-party liabilities, for the exceptions listed in Clause 22(2)).

Clauses 21 and 23 provide safeguards for the Employer by ensuring, through the insurances that the Contractor is required to provide (up to the limits of indemnity stated in the policies), that the Contractor will be able to meet any obligations that arise and for which he is responsible under Clauses 20 and 22.

The Contractor may have a limit of indemnity on his third-party policy that is in excess of the limit stated in the Contract. This would provide the Contractor (and therefore the Employer) with additional protection because the Contractor's liability is not limited to the minimum amount stated in the Contract. However, the Contractor in all cases remains liable for the total amount should any claim be settled at an amount in excess of the minimum amount of insurance cover required under the Contract (Clauses 21(1) and 23(3)).

The Contractor's responsibility to carry employer's liability insurance is a statutory obligation and thus needs no provision in the ICE Conditions of Contract.

No reference is made to professional indemnity insurance for any design work required under the Contract. Design of Temporary Works that is incidental to work being carried out under the Contract is normally covered under the policy provided for the insurance of the Works (Clause 21). Design liability for any part of the Permanent Works, if it is to be covered by insurance, would have to be covered under a separate special professional indemnity insurance policy agreed by the Parties (see note against Clause 8(2)).

Clause 20: Care of the Works

Under Clause 20(1)(a) the Contractor is made fully responsible for the care of the Works and for all materials, plant and equipment for incorporation therein from the Works Commencement Date (defined in Clause 41(1)), normally until substantial completion.

Under Clause 20(1)(a) the Employer takes over full responsibility for completed work from the date of issue of a Certificate of Substantial Completion. It is therefore essential for the Engineer to give the Employer advance notice of his intention to certify completion so that any requisite insurance can be arranged.

Clause 21(1): Insurance of the Works etc.

The Contractor's insurance policy covering the Works must be for the full replacement cost plus an additional 10% to cover additional costs that may arise incidental to the rectification of any loss or damage, but there is no requirement under the Contract for the Contractor to insure his own plant and equipment.

The Contractor may, for his own protection, have some insurance cover for his own plant and equipment against loss through accident, fire, theft and the like.

Clause 25(1): Evidence and terms of insurance

It is in the Employer's interests that insurance policies taken out on his behalf are verified as meeting the requirements of the Contract and that there is evidence of payment of premiums.

Clause 25(2): Excesses

The excesses carried by a Contractor on his insurance policies must be disclosed at the time of tender by his entering the relevant amounts in Part 2 of the Appendix to the Form of Tender. The excesses (or uninsured losses) that the Contractor carries under his insurance policies are the Contractor's own responsibility. Although it should be verified that the excesses quoted are not too high for a particular project, it would not be appropriate for them to be queried unless they appear to be unduly large in relation to the known resources of the Contractor.

Clause 26: Giving of notices and payment of fees

Unless indicated otherwise in the Contract, all notices will be given by the Contractor and any fees required will be paid by the Contractor with the cost being reimbursed under Clause 26(2).

It is not intended that the Contractor should be responsible for obtaining any planning permission in respect of the Permanent Works or for any Temporary Works designed by the Engineer in their designated positions on Site.

If any planning permission as required under sub-clause (3)(c) has not been obtained prior to commencement of work on Site or in due time the warranty given ensures that the Contractor will be granted an extension of time for completion of the Works and be reimbursed for any additional costs incurred should there be any delay in the granting of the planning permission required.

Clause 27: New Roads and Street Works Act, 1991

The Contractor must be given full details of all conditions or limitations imposed by the highway authority or by any other responsible authority when a licence is granted or consent is given for the carrying out of street works covered by the Act. Other authorities may include but are not limited to traffic, bridge, sewer and transport authorities.

Statutory obligations arising under the New Roads and Street Works Act 1991 may extend beyond the date of issue of the Defects Correction Certificate.

Clause 36: Quality of materials and workmanship and tests

Quality Assurance procedures are not referred to as such in the Seventh Edition of the ICE Conditions of Contract. It should be made clear in the tender documentation if they are required and details should be given in the Specification.

Clause 41: Works Commencement Date

The Works Commencement Date is defined as one of three alternatives and is the starting date for the time allowed for the completion of the Contract, irrespective of when work actually commences.

Clause 43: Time for completion

Either the Employer or the Contractor is required to state in the Appendix to the Form of Tender the completion time for the whole of the Works or completion times for each of the separately identified Sections comprising the whole of the Works.

Where completion times are stated for separately identified Sections of the Works it is necessary to ensure that the Sections so identified plus "the Remainder of the Works" cover the whole of the Works.

Clause 44: Extension of time for completion

Under Clause 44(2) the Engineer must make an assessment of the delay suffered by the Contractor upon receipt of any particulars of delay under Clause 44(1). In addition, in the absence of any notification, the Engineer may make a similar assessment if he considers that the Contractor has in fact suffered a delay. In either event the Contractor must be notified in writing of the Engineer's assessment.

This requirement is included to ensure the timely recording by the Engineer of any delay suffered or claimed to have been suffered by the Contractor, but there is no need for any decision to be made at that time as to whether any extension of time for completion of the Works should be granted.

Under Clause 44(3) the Engineer has a duty to grant an interim extension of time for completion as soon as he considers that such an extension is due. This is to ensure that the Contractor does not incur additional cost by accelerating progress unnecessarily due to doubt as to whether or not an extension of time will be granted. If a claim for an extension of time is rejected the Contractor must be notified without delay.

Clause 47: Liquidated damages for delay in substantial completion

Clause 47 covers in detail the different situations that can exist either in dealing with the whole of the Works (Clause 47(1)) or where the Works is divided into Sections (Clause 47(2)). There is also provision for a reduction in the damages payable should any part of the Works or part of a Section of the Works be certified as complete.

When the Works is divided into Sections the Appendix to the Form of Tender states the Liquidated Damages payable should completion of any of the separately identified Sections of the Works (including the Remainder of the Works) be delayed.

It must be noted that the liquidated damages for delay quoted in the Appendix to the Form of Tender are alternatives and that if completion dates are quoted or required for Sections then liquidated damages should not be quoted for failure to complete the whole of the Works within the overall time for completion.

Provision is made in Clause 47(4)(a) for a limit to be placed on liquidated damages if this is thought to be appropriate.

Clause 47(4)(b) provides that if no sum or sums for liquidated damages are quoted in the Appendix to the Form of Tender then no damages, whether liquidated or unliquidated, will be payable.

Clause 47(6) sets out the procedure to be followed if, after liquidated damages have become payable, the Engineer issues a variation order or accepts that a Clause 12 situation or any other situation outside the control of the Contractor exists which in either event will involve further delay to that part of the Works.

In such a case the Employer's entitlement to liquidated damages is suspended until the further delay comes to an end. Both the commencement and termination of such further delay must be notified in writing by the Engineer.

Clause 49 : Work outstanding

It should be noted that the ICE Conditions of Contract unamended are not suitable for landscaping work or tree planting etc. for which an extended period of care and maintenance is usually required.

If a Contract let under the ICE Conditions of Contract does include an element of landscaping work or tree planting it must be made clear that any extended care and maintenance requirements that have to continue beyond the end of the Defects Correction Period will not be considered as defects to be remedied before the issue of the Defects Correction Certificate. Payment of the balance of the retention money, due to be paid under Clause 60(6), will therefore not be delayed on account of such outstanding work.

Clauses 51(1) and 52(1): Ordered variations and their valuation

Clause 51(1) specifically permits the ordering of variations during the Defects Correction Period. In the valuation of such variations Clause 52(1)(b) permits account to be taken for example of the fact that by then the Contractor may well have left the Site and may need to return to it specially.

Clause 52 makes it clear that all efforts should be made to agree the value of a variation and the delay consequences before the varied work is carried out.

Clause 53: Additional Payments

The procedure for claiming additional payments under the Contract is set out in detail in Clause 53. The Contractor's right to payment is subject to him providing details of his claim to the satisfaction of the Engineer.

Clause 54: Non-removal of Materials and Contractor's Equipment

It should be noted that there is no provision for goods or materials owned by the Contractor becoming, when on Site, the property of the Employer.

Clause 59: Nominated Sub-contractors

Whether or not Nominated Sub-contractors are to be used is a choice made by the Employer, on the advice of the Engineer. In many cases it may be possible to select the Nominated Sub-contractor during the pre-tender stage of the procurement thus allowing tenderers to take the work required fully into account.

Provision is made under Clause 59(1) for the Contractor to be able to object to the employment of a particular Nominated Sub-contractor against whom he has a reasonable objection or who declines to enter into a sub-contract containing provisions accepting obligations generally similar to those undertaken by the Contractor.

Particular note should be taken of Clause 59(1)(d) under which the Nominated Sub-contractor can be required to provide a performance security that is acceptable to the Contractor, and also of the procedure set out in Clause 59(4) that is to be followed in the event of a Nominated Sub-contractor's default.

In any case of a Nominated Sub-contractor's default the Contractor must notify the Engineer in writing before any action can be taken. The first option open to the Contractor, with the written consent of the Engineer, is to terminate the Sub-contractor's employment and to recover damages from the Sub-contractor or under the security provided by the Sub-contractor. The Engineer must then take action under Clause 59(2) and renominate or take such other action as he deems to be appropriate.

If the Engineer, because of design or other considerations, does not agree to the termination of that Nominated Sub-contractor's employment Clause 59(4)(b) provides that the Engineer must issue appropriate instructions under Clause 13.

When a sub-contractor has been nominated it should be borne in mind that the Employer may in certain circumstances have a residual responsibility if the Contractor is involved in additional costs and delays which he could not have reasonably foreseen and which may therefore entitle him to reimbursement under the Contract.

Clause 60: Payment Provisions

The payment provisions now take account of the requirements for the Housing Grants, Construction and Regeneration Act 1996 with respect to timing and the provision of information.

There is no change in the traditional procedures for payment against interim and final certificates.

Clause 60(2): Monthly payments

This Clause maintains the requirement that the final date for payment by the Employer against an Engineer's Certificate is 28 days after the date of delivery of the Contractor's monthly statement to the Engineer or the Engineer's Representative.

There is however an additional requirement whereby the Engineer has to issue and deliver his Certificate to the Employer (with a copy to the Contractor) within 25 days of the date of delivery of the Contractor's monthly statement to the Engineer's Representative.

Payment to the Contractor becomes due on certification by the Engineer with the final date for payment being within 28 days of delivery of the monthly statement to the Engineer or the Engineer's Representative.

Clause 60(5): Retention

The rate of retention to be deducted in accordance with Clause 60(5) is to be stated in the Appendix to the Form of Tender, Part 1 and it is recommended that this should not exceed 5% . The limit of retention has also to be stated in the Appendix to the Form of Tender, Part 1 and it is recommended that this should not exceed 3% of the Tender Total.

Clause 60(7): Interest on overdue payments

Clause 60(7) states that interest is due to be paid for each day a payment is overdue and that the interest is to be compounded monthly.

As the Engineer is required to certify on the basis of the statement submitted by the Contractor under Clause 60(1) the Contractor when submitting each monthly statement should include the amount of all interest that he claims as due under Clause 60(7) up to that statement date in his monthly valuation under Clause 60(1)(d).

The bank whose base lending rate is to be used for the calculation of interest must be specified in the Appendix to the Form of Tender.

Clause 60(7) makes provision for the payment of interest on sums payable to the Contractor after certificates have been corrected by an arbitrator.

Clause 60(9): Certificates

This Clause provides that the Engineer's certificate shall also serve as the Employer's notification to the Contractor of the amount to be paid and the basis on which it was calculated. As the Engineer's certificate has to be issued at least 3 days before the final date for payment the Contractor has at least 3 days notice of the reduction made by the Engineer from the amount claimed by the Contractor in his statement.

If the information is not provided with the Engineer's certificate the Contractor is entitled to ask for full details as to how the amount certified by the Engineer has been arrived at. This information could be essential to enable the Contractor to make payments at the time they are due to any sub-contractors employed on the Contract. Details of any alterations in the rates claimed by the Contractor for new or changed items or any reduction in the quantities claimed could also be necessary so that the Contractor can decide whether to indicate his dissatisfaction with the valuation and the certificate issued under the provisions of Clause 66(2).

Clause 60(10): Payment advice

This Clause supplements the provisions of Clause 60(9) and requires that the Contractor be notified by the Employer at least one day before a payment is due if the payment being made will differ in any respect from the amount certified by the Engineer. Details of how any amount being withheld has been calculated and the grounds for so doing have to be stated.

Clause 64: Default of the Employer

This Clause makes provision if the Employer is in default and sets out the events upon which the Contractor may rely to terminate his employment under the Contract after giving due notice to the Employer. This Clause balances the provisions of Clause 65 which covers any default of the Contractor.

CLAUSE 66: AVOIDANCE AND SETTLEMENT OF DISPUTES

This Clause is drafted to comply with the adjudication provisions of the Housing Grants, Construction and Regeneration Act 1996. The intention is that the adjudication procedure provided for by the Act's associated "Scheme for Construction Contracts" should not apply.

Clause 66(1): Avoidance of disputes

The disputes procedure now includes options for both conciliation and adjudication before there is a reference to arbitration. The intention behind Clause 66 is to overcome where possible the causes of disputes and, in those cases where disputes are still likely to arise, to facilitate their clear definition and early resolution preferably by agreement. It is possible there could be consequential savings in both time and cost.

Clause 66(2): Matters of dissatisfaction

If at any time the Employer or the Contractor is dissatisfied with any decision opinion instruction direction certificate or valuation of the Engineer or with any other matter in connection with or arising out of the Contract or the carrying out of the Works the matter has to be referred to and decided by the Engineer.

The Engineer has to give his decision on all matters included in the written submission of the referring party within one calendar month after receiving that written submission. This decision of the Engineer is an essential preliminary before the Disputes procedure set out in Clause 66 can come into force.

If either party is dissatisfied with the Engineer's decision or if a decision has not been given within the time allowed either party may serve a Notice of Dispute.

Clause 66(3): Disputes

This Clause sets out what can give rise to or constitute a dispute or difference leading to a Notice of Dispute.

Settlement of disputes
Clause 66(5): Conciliation
Clause 66(6): Adjudication
Clause 66(9): Arbitration

For the settlement of disputes there are options for conciliation or adjudication before there is a reference to arbitration. It is anticipated that most disputes will be dealt with as soon as possible after they arise. With the potential for the prompt settlement of many minor disputes or differences that can arise on a Contract this should lead to considerable savings in time and cost.

Clause 66(5): Conciliation

After service of a Notice of Dispute the dispute can be considered under the provisions of the Institution of Civil Engineers Conciliation Procedure 1994 (see Clause 66(5)) or, after Notice of Adjudication has been given (see Clause 66(6)) may be referred to an adjudicator under the terms of the Contract. These two alternatives provide routes for a speedy settlement of most disputes.

If the recommendation of a conciliator is not accepted by either party within the laid down time limit (one month) the dispute may be referred to Adjudication or direct to Arbitration. If the decision of an Adjudicator is not accepted by either party the dispute may be referred to Arbitration within the laid down time limit (three months).

Clause 66(6): Adjudication

The Contract incorporates by reference the ICE Adjudication Procedure 1997 (or any amendment or modification in force at the time of the Notice of Adjudication).

Following a Notice of Adjudication, 7 days are allowed for appointing the adjudicator and referring the dispute to him, whereupon he is given 28 days (or such other period as may be permitted) in which to reach his decision.

Once a decision has been reached and notified to the parties it becomes binding unless and until the dispute (not the decision) is referred to arbitration.

A failure to give effect to a decision is excluded from the arbitration agreement leaving the parties free to seek enforcement through the courts.

If a dispute on which a decision has been given is not referred to arbitration by issuing a Notice to Refer under Clause 66(9) within three months of the giving of the decision, that decision becomes final as well as binding and cannot thereafter be challenged.

Clause 67(3): Application to Northern Ireland

The provisions of the Housing Grants, Construction and Regeneration Act 1996 (HGCR Act 1996) have been covered by am
in these Conditions of Contract but when the Construction Contracts (Northern Ireland) Order 1997 (CC(NI) Order 1997) was published it had a different clause numbering system from that used in the HGCR Act 1996.

For ready reference the Sections of the HGCR Act 1996 are cross referenced below to the applicable Paragraphs of the CC(NI) Order 1997. It should however be noted that there are differences in the text between the Act and the Order. The HGCR Act 1996 came into force on 1 May 1998. The CC(NI) Order 1997 came into force on 1 June 1999.

HGCR Act 1996 Section	Construction Contracts (NI) Order Paragraph
104	3
105	4
106	5
107	6
108	7
109	8
110	9
111	10
112	11
113	12
114	13 and 16
115	14
116	14(c)
117	15

Clause 69(1)(b): Labour and landfill tax fluctuation

The new Clause 69 applies also to the Finance Act 1996 (Sections 39-71 and Schedule 5) and the Landfill Tax Regulations 1996. It is made clear that only those rates of Landfill Tax chargeable under the above Act and Regulations in force at the date for return of tenders shall be deemed to have been included in the tender rates, whether or not future increases or decreases are then foreseeable. Increases or decreases after the date for return of tenders are covered in Clause 69(2).

Clause 71: CDM Regulations 1994

The Construction (Design and Management) Regulations apply to all construction contracts, other than:

- Site surveys;
- Exploration and abstraction of mineral resources and activities preparatory thereto carried out at the site of such exploration or abstraction;
- Projects for which the construction phase will not last more than 30 days and which do not involve more than 500 man days of construction;
- Projects for which the local authority is the enforcing authority under the Health and Safety (Enforcing Authority) Regulations 1989;
- Construction (but not demolition) where not more than 5 persons are employed at any one time; or
- Construction for a private client in respect of his own dwelling.

Anybody entering into a contract where the CDM Regulations apply should study the Regulations in detail. The list of exclusions above is not exhaustive.

Under these Regulations the Employer is required to appoint, as soon as practicable, a PLANNING SUPERVISOR responsible, among other things, for the preparation and monitoring of a HEALTH AND SAFETY PLAN for the contract, or for the project of which the contract is a part.

The name of the Planning Supervisor is included in Part 1 of the Appendix to the Form of Tender, and the Health and Safety Plan for the Works and the Health and Safety Files in respect of each structure included in the Works (to the extent that they have been prepared) will be available for inspection by tenderers.

Under these Regulations the Employer is also required to appoint for the project as soon as practicable a PRINCIPAL CONTRACTOR with particular responsibilities regarding safety for the project whose name (if one has been appointed by that time) is also included in Part 1 of the Appendix to the Form of Tender.

The Contractor could also be asked to act as Principal Contractor and may also be asked to act as Planning Supervisor.

On completion of the Contractor's obligations under the Contract, the Safety File becomes the property of and must be handed over to the Employer.

Where these responsibilities are not covered by separate contracts they are likely to be covered in the Specification, with appropriate items inserted in the Bill of Quantities.

Clause 72: Special conditions

As indicated in the section on Contract Documents (page 3), special conditions may be added

Contract Price Fluctuation Clauses

Contract Price Fluctuation Clauses are printed in the Seventh Edition and are also supplied in loose-leaf form. When one (or more) is to be incorporated in a Contract it should be included as a special Condition.

If a Contract Price Fluctuation Clause is to be included in a Contract attention is drawn to Appendix 2 of these Guidance Notes. This Appendix sets out the views of the CCSJC on what constitutes good practice when a Contract Price Fluctuation Clause is used.

APPENDIX 1

PERFORMANCE BOND

General The Bond may be called only if the Contractor is expelled from the Site as a result of insolvency or is removed from the Site under any other of the provisions of Clause 65. If the Contractor has not been expelled from the Works, there is sufficient redress in the Contract (e.g. retention, set-off, damages etc.) to deal with other matters which might constitute a default. The Bond relates to Clause 65 for matters of expulsion and certification subsequent to that expulsion, and to the Contractor's failure to pay the Excess Sum (or any part of the Excess Sum remaining unpaid).

Surety Bond The reference to Clause 66 enables the Surety, if he so wishes, to utilise the same adjudication provisions which are available to the Contractor in order to challenge the Engineer's Certificate.

Although the Surety has the right to receive copies of any notice served on the Contractor under Clause 65(1) or any certificate served under Clause 65(4) or 65(5), and to request information or to inspect the Site, this should not be considered to be a condition precedent to payment. The only condition precedent to payment by the Surety is the service on the Surety of the **certified copy** Clause 65(5) certificate. This certificate must be certified by the Engineer as a true copy and it may be an interim certificate or a final one.

By accepting this bond, the Surety undertakes, within 14 days of receiving the certified copy of a Clause 65(5) certificate, to either pay or refer the matter to adjudication. If the Surety wishes to contest the certificate or if the Employer refuses to e.g. allow the Surety's representative to inspect the Site or provide information which the Surety requests, he must seek the decision of the adjudicator in the first instance under the provisions of Clauses 66(6) to 66(8) of the Contract. The adjudicator's decision is binding unless and until the dispute is finally determined by legal proceedings, by arbitration or by agreement.

The provision in clause 1(6) of the Bond in relation to a previous adjudication between the Employer and the Contractor prevents a second adjudication taking place on the same matter. However a safeguard is incorporated to the effect that both parties must have made submissions to the adjudicator before his decision was given. This is necessary in the event that the Contractor is unable to respond to an adjudication.

The time periods have been set so that payment is not delayed unnecessarily in the event of there being no dispute. On the other hand, in the event of a dispute, the Surety has sufficient time to protect his interests. Clause 1(7) of the Bond specifies the time for payment at 7 days from the decision of the adjudicator in the event that there is no other direction.

APPENDIX 2

CONTRACT PRICE FLUCTUATION

Foreword Three Contract Price Fluctuation Clauses are printed in the Seventh Edition of the ICE Conditions of Contract for inclusion in a Contract as special conditions when required. Loose leaf copies are also available. The Index Figures used are compiled by the Department of the Environment Transport and the Regions and published by The Stationary Office in the Monthly Bulletin of Indices "1990 Series Civil Engineering Formula Indices" and "1990 Series Structural Steelwork Formula Indices".

The CEW Clause is provided for use in connection with Works of Civil Engineering Construction, the SS Clause is provided for use on Contracts that are wholly or predominantly in respect of Structural Steelwork and the CEW/SS Clause has to be incorporated with both the CEW Clause and the SS Clause when provision is to be made for the adjustment of price fluctuation on Contracts involving both Civil Engineering Work and Structural Steelwork.

In the Guidance Notes that follow, reference to the Contract Price Fluctuations Clause (CPF Clause) is to be taken to refer to the CEW Clause or to the SS Clause or to the CEW, the SS and the CEW/SS Clauses.

1. Introduction The CPF Clause offers valuable savings in administrative effort compared with earlier variation of price clauses. These savings are achieved because the complexities arising from the effects of price variations on a contract are reduced to one simple formula. To achieve this degree of simplification many approximations and generalizations must be made: for example, the Engineer's assessment of the proportions in which different types of work contribute to the Works as a whole and the assumption that these different types of work will be performed at a steady rate throughout the period of the Contract.

It is necessary for both the Employer and the Contractor to accept that degree of approximation which is inherent in the CPF Clause. It is equally important that particular aspects of the CPF Clause should not be viewed in isolation as being likely to produce advantages or disadvantages to either the Employer or the Contractor.

Where contracts are awarded by competitive tendering any advantages or disadvantages which tenderers foresee arising from the approximations inherent in the CPF Clause will be reflected in their tenders and are unlikely to give rise to inequitable payments.

It is desirable that a uniform method of operating the CPF Clause should be established to enable tenderers to base their tenders on uniform precepts. The CCSJC is unanimous that in most contracts the methods of operating the CPF Clause set out in sections 2-4 of this Appendix are good practice and in accordance with business sense.

If, in an exceptional case, the CPF Clause is to be operated in a different manner, the proposed alternative should be described in the tender documents to enable tenderers to take its effects fully into account in the Contract.

2. Retention The CCSJC takes the view that the contractual entitlement to any payment is dependent upon Clause 60(2) and that the CPF Clause only provides for a method of calculating the fluctuation payments due as part of that payment. Payments due under the CPF Clause will be shown by the Contractor in his monthly statement in accordance with Clause 60(1)(d). Appropriate payments will be certified by the Engineer in accordance with Clause 60(2)(a) and will thus be subject to retention in accordance with Clause 60(5).

3. Dayworks and Nominated Sub-contractors

The CCSJC considers that the words "... based on actual cost or current prices ..." in CEW Clause (2)(d)(i) apply to both Dayworks and Nominated Sub-contractors. Thus, amounts for Dayworks, or parts thereof, and Nominated Sub-contractors, where paid at actual cost or current prices, should be excluded from the Effective Value while those amounts based on rates and prices quoted at the time of tender should be included in the Effective Value.

4. Goods and materials on Site or vested in the Employer

The amounts included in an interim certificate pursuant to Clause 60(2)(b) in respect of goods or materials on Site or vested in the Employer are normally a percentage (in no case exceeding the percentage stated in the Appendix to the Form of Tender) of the invoiced prices of the goods or materials.

When calculating the Effective Value, the above amounts, because they are based on the invoiced prices and thus items based on actual cost, should be deducted under CEW Clause (2)(d)(i) from the amount which, in the opinion of the Engineer, would otherwise be due to the Contractor under CEW Clause 2(d)(i), i.e. they do not at that stage form part of the Effective Value at that time.

In the month in which any of these goods or materials are incorporated into the Permanent Works the value of those goods or materials at tender prices (being an element in rates in the Bill of Quantities) will form part of the estimated contract value of the Permanent Works certified under Clause 60(2)(a) and thus become part of the Effective Value at that time. The Contractor will thus be paid a monthly amount in respect of the tender value of the goods or materials under Clause 60(2)(a) plus a fluctuation payment under CEW Clause (2)(d)(i) increasing or decreasing the tender value of the goods or materials in line with the increase or decrease in the value of the Works as a whole as reflected through the Price Fluctuation Factor. The percentage of the invoiced price previously paid will effectively be deducted as a previous payment on account through the operation of Clause 60(2).

5. Comment

Attention has been directed to the case where, during a period of high inflation, goods and materials remain on Site for a considerable time, during which the Price Fluctuation Factor increases significantly. Although, as stated in section 1, particular aspects of the CPF Clause should not be viewed in isolation the CCSJC considers that this matter requires special mention. It has been suggested that in the circumstances described the Contractor would be paid an unjustifiably higher price for goods and materials than he need have been paid had the goods and materials been subject to the Price Fluctuation Factor of the month in which the goods or materials were included in an interim certificate.

However, when goods or materials are brought on to Site expediently it contributes to timely completion and for this added safeguard the Employer has paid only a percentage of the invoiced price of the goods or materials in advance of their incorporation into the Works. The longer the goods or materials remain on Site the longer is the Contractor obliged to finance that percentage of the goods or materials which has not been paid for by the Employer. The Contractor is also responsible for replacing any goods or materials stolen or spoilt while on Site. Viewed in the context of the CPF Clause the CCSJC considers that neither party receives inequitable treatment.

ICE Conditions of Contract

Seventh Edition

CONDITIONS OF CONTRACT AND FORMS OF TENDER, AGREEMENT AND BOND FOR USE IN CONNECTION WITH WORKS OF CIVIL ENGINEERING CONSTRUCTION

Measurement Version

Institution of Civil Engineers
Association of Consulting Engineers
Civil Engineering Contractors Association

Published by Thomas Telford Publishing, Thomas Telford Limited, 1 Heron Quay, London E14 4JD on behalf of:

The Director General
Institution of Civil Engineers
One Great George Street
London SW1P 3AA

The Chief Executive
Association of Consulting Engineers
Alliance House
12 Caxton Street
London SWlH 0QL

The Director
Civil Engineering Contractors Association
Construction House
56-64 Leonard Street
London EC2A 4JX

First edition was agreed by the Institution of Civil Engineers and the Federation of Civil Engineering Contractors, December 1945
Second to Sixth editions were approved by the Institution of Civil Engineers, the
Association of Consulting Engineers and the Federation of Civil Engineering Contractors
Second edition, January 1950
Third edition, March 1951
Fourth edition, January 1955
Fifth edition, June 1973; revised January 1979; reprinted with amendments January 1986
Sixth edition, January 1991; reprinted with amendments November 1995;
reprinted with amendments November 1997
Seventh edition was approved by the Institution of Civil Engineers, the Association of Consulting Engineers and the Civil Engineering Contractors Association, September 1999.
reprinted with amendments January 2003

The Institution of Civil Engineers, Association of Consulting Engineers and Civil Engineering Contractors Association have, as sponsoring authorities, approved this revised, seventh, edition of the document commonly known as the ICE Conditions of Contract, for all works of civil engineering construction. A permanent joint committee has prepared and will keep under review the use of the document and will consider any suggestions for amendment, which should be addressed to the Director General (CCSJC), Institution of Civil Engineers, One Great George Street, London SWlP 3AA. Revision to the document will be made when such action seems warranted.

9 8 7 6

ISBN 0 7277 2789 3

A catalogue record for this book is available from the British Library

Printed and bound in Great Britain by Formara Ltd, Essex.

CONTENTS

Associated publications referred to in the Contract

Guidance Notes to the ICE Conditions of Contract 7th Edition
ICE Arbitration Procedure 1997
ICE Appendix 2001 to the Scottish Arbitration Code 1999
ICE Conciliation Procedure 1999
ICE Adjudication Procedure 1997

Other ICE Conditions of Contract publications

Guidance Notes to the ICE Conditions of Contract Measurement Version 7th Edition
ICE Design and Construct Conditions of Contract 2nd Edition
Guidance Notes to the ICE Design and Construct Conditions of Contract 2nd Edition
ICE Conditions of Contract Minor Works 3rd Edition
ICE Conditions of Contract Term Version 1st Edition
Guidance Notes to the ICE Conditions of Contract Term Version 1st Edition
ICE Conditions of Contract for Ground Investigation
Tendering for Civil Engineering Contracts
Agreement for Consultancy Work in Respect of Domestic or Small Works

ACKNOWLEDGEMENTS

The members of the Conditions of Contract Standing Joint Committee are:

D. H Hodgkinson (Chairman) MSc CEng FICE MIHT
G. F. Hawker (Legal Advisor) TD BSc(Eng) FEng FICE FIEI
I. Elliott BSc CEng MICE MCIWEM
W. R. Woolley CEng MICE
A. Scott
J. G. Rees BSc(Eng) FICE FCIArb
D. Tadiello
A. Kondys FRICS
R. Beales
R. D. Nokes

Secretariat
F. N. Vernon (Technical Adviser) BSc, CEng, MICE
J. M. Hawkins BA (Hons)

CONTENTS OF THE ICE CONDITIONS OF CONTRACT

MATERIALS AND WORKMANSHIP

COMMENCEMENT TIME AND DELAYS

LIQUIDATED DAMAGES FOR DELAY

CERTIFICATE OF SUBSTANTIAL COMPLETION

OUTSTANDING WORK AND DEFECTS

ALTERATIONS ADDITIONS AND OMISSIONS

PROPERTY IN MATERIALS AND CONTRACTOR'S EQUIPMENT

MEASUREMENT

PROVISIONAL AND PRIME COST SUMS AND NOMINATED SUB-CONTRACTS

CERTIFICATES AND PAYMENT

INDEX

ICE Conditions of Contract - Measurement Version

DEFINITIONS AND INTERPRETATION

Definitions 1 (1) In the Contract (as hereinafter defined) the following words and expressions shall have the meanings hereby assigned to them except where the context otherwise requires.

(a) "Employer" means the person or persons firm company or other body named in the Appendix to the Form of Tender and includes the Employer's personal representatives successors and permitted assignees.

(b) "Contractor" means the person or persons firm or company to whom the Contract has been awarded by the Employer and includes the Contractor's personal representatives successors and permitted assignees.

(c) "Engineer" means the person firm or company appointed by the Employer to act as Engineer for the purposes of the Contract and named in the Appendix to the Form of Tender or any other person firm or company so appointed from time to time by the Employer and notified in writing as such to the Contractor.

(d) "Engineer's Representative" means a person notified as such from time to time by the Engineer under Clause 2(3)(a).

(e) "Contract" means the Conditions of Contract Specification Drawings Bill of Quantities the Form of Tender the written acceptance thereof and the Form of Agreement (if completed).

(f) "Specification" means the specification referred to in the Form of Tender and any modification thereof or addition thereto as may from time to time be furnished or approved in writing by the Engineer.

(g) "Drawings" means the drawings referred to in the Specification and any modification of such drawings approved in writing by the Engineer and such other drawings as may from time to time be furnished by or approved in writing by the Engineer.

(h) "Bill of Quantities" means the priced and completed Bill of Quantities.

(i) "Tender Total" means the total of the Bill of Quantities at the date of award of the Contract or in the absence of a Bill of Quantities the agreed estimated total value of the Works at that date.

(j) "Contract Price" means the sum to be ascertained and paid in accordance with the provisions hereinafter contained for the construction and completion of the Works in accordance with the Contract.

(k) "Prime Cost (PC) Item" means an item in the Contract which contains (either wholly or in part) a sum referred to as Prime Cost (PC) which will be used for the carrying out of work or the supply of goods materials or services for the Works.

(l) "Provisional Sum" means a sum included and so designated in the Contract as a specific contingency for the carrying out of work or the supply of goods materials or services which may be used in whole or in part or not at all at the direction and discretion of the Engineer.

(m) "Nominated Sub-contractor" means any merchant tradesman specialist or other person firm or company nominated in accordance with the Contract to be employed by the Contractor for the carrying out of work or supply of goods materials or services for which a Prime Cost or a Provisional Sum has been included in the Contract.

(n) "Permanent Works" means the permanent works to be constructed and completed in accordance with the Contract.

(o) "Temporary Works" means all temporary works of every kind required in or about the construction and completion of the Works.

(p) "Works" means the Permanent Works together with the Temporary Works.

(q) "Works Commencement Date"—as defined in Clause 41 (1).

(r) "Certificate of Substantial Completion" means a certificate issued under Clause 48.

(s) "Defects Correction Period" means that period stated in the Appendix to the Form of Tender calculated from the date on which the Contractor becomes entitled to a Certificate of Substantial Completion for the Works or any Section or part thereof.

(t) "Defects Correction Certificate"—as defined in Clause 61 (1).

(u) "Section" means a part of the Works separately identified in the Appendix to the Form of Tender.

(v) "Site" means the lands and other places on under in or through which the Works are to be constructed and any other lands or places provided by the Employer for the purposes of the Contract together with such other places as may be designated in the Contract or subsequently agreed by the Engineer as forming part of the Site.

(w) "Contractor's Equipment" means all appliances or things of whatsoever nature required in or about the construction and completion of the Works but does not include materials or other things intended to form or forming part of the Permanent Works.

Singular and plural (2) Words importing the singular also include the plural and vice-versa where the context requires.

Headings and marginal notes (3) The headings and marginal notes in the Conditions of Contract shall not be deemed to be part thereof or be taken into consideration in the interpretation or construction thereof or of the Contract.

Clause references (4) All references herein to clauses are references to clauses numbered in the Conditions of Contract and not to those in any other document forming part of the Contract.

Cost (5) The word "cost" when used in the Conditions of Contract means all expenditure properly incurred or to be incurred whether on or off the Site including overhead finance and other charges properly allocatable thereto but does not include any allowance for profit.

Communications in writing (6) Communications which under the Contract are required to be "in writing" may be hand-written typewritten or printed and sent by hand post telex cable facsimile or other means resulting in a permanent record.

ENGINEER AND ENGINEER'S REPRESENTATIVE

Duties and authority of Engineer

2 (1) (a) The Engineer shall carry out the duties specified in or necessarily to be implied from the Contract.

(b) The Engineer may exercise the authority specified in or necessarily to be implied from the Contract. If the Engineer is required under the terms of his appointment by the Employer to obtain the specific approval of the Employer before exercising any such authority particulars of such requirements shall be those set out in the Appendix to the Form of Tender. Any requisite approval shall be deemed to have been given by the Employer for any such authority exercised by the Engineer.

(c) Except as expressly stated in the Contract the Engineer shall have no authority to amend the Contract nor to relieve the Contractor of any of his obligations under the Contract.

(d) The giving of any consent or approval by or on behalf of the Engineer shall not in any way relieve the Contractor of any of his obligations under the Contract or of his duty to ensure the correctness or accuracy of the matter or thing which is the subject of the consent or approval.

Named individual

(2) (a) Where the Engineer as defined in Clause 1 (1)(c) is not a single named Chartered Engineer the Engineer shall within 7 days of the award of the Contract and in any event before the Works Commencement Date notify to the Contractor in writing the name of the Chartered Engineer who will act on his behalf and assume the full responsibilities of the Engineer under the Contract.

(b) The Engineer shall thereafter in like manner notify the Contractor of any replacement of the named Chartered Engineer.

Engineer's Representative

(3) (a) The Engineer's Representative shall be responsible to the Engineer who shall notify his appointment to the Contractor in writing.

(b) The Engineer's Representative shall watch and supervise the construction and completion of the Works. He shall have no authority

(i) to relieve the Contractor of any of his duties or obligations under the Contract

nor except as expressly provided for in sub-clause (4) of this Clause

(ii) to order any work involving delay or any extra payment by the Employer or

(iii) to make any variation of or in the Works.

Delegation by Engineer

(4) The Engineer may from time to time delegate to the Engineer's Representative or any other person responsible to the Engineer any of the duties and authorities vested in the Engineer and he may at any time revoke such delegation. Any such delegation

(a) shall be in writing and shall not take effect until such time as a copy thereof has been delivered to the Contractor or his agent appointed under Clause 15(2)

(b) shall continue in force until such time as the Engineer shall notify the Contractor in writing that the same has been revoked

(c) shall not be given in respect of any decision to be taken or certificate to be issued under Clauses 12(6) 44 48 60(4) 61 65 or 66.

Assistants

(5) (a) The Engineer or the Engineer's Representative may appoint any number of persons to assist the Engineer's Representative in the carrying out of his duties under sub-clause (3)(b) or (4) of this Clause. He shall notify to the Contractor the names duties and scope of authority of such persons.

(b) Such assistants shall have no authority to issue any instructions to the Contractor save insofar as such instructions may be necessary to enable them to carry out their duties and to secure the acceptance of materials and workmanship as being in accordance with the Contract. Any instructions given by an assistant for these purposes shall where appropriate be in writing and be deemed to have been given by the Engineer's Representative.

(c) If the Contractor is dissatisfied by reason of any instruction of any assistant of the Engineer's Representative appointed under sub-clause (5)(a) of this Clause he shall be entitled to refer the matter to the Engineer's Representative who shall thereupon confirm reverse or vary such instruction.

Instructions

(6) (a) Instructions given by the Engineer or by any person exercising delegated duties and authorities under sub-clause (4) of this Clause shall be in writing. Provided that if for any reason it is considered necessary to give any such instruction orally the Contractor shall comply therewith.

(b) Any such oral instruction shall be confirmed in writing as soon as is possible under the circumstances. Provided that if the Contractor confirms in writing any such oral instruction which confirmation is not contradicted in writing by the Engineer or the Engineer's Representative forthwith it shall be deemed to be an instruction in writing by the Engineer.

(c) Upon the written request of the Contractor the Engineer or the person exercising delegated duties or authorities under sub-clause (4) of this Clause shall specify in writing under which of his duties and authorities the instruction is given.

Impartiality

(7) The Engineer shall act impartially within the terms of the Contract having regard to all the circumstances. In like manner the Engineer's Representative and any person exercising delegated duties and authorities shall also act impartially.

ASSIGNMENT AND SUB-CONTRACTING

Assignment 3

(1) Neither the Employer nor the Contractor shall assign the Contract or any part thereof or any benefit or interest therein or thereunder without the prior written consent of the other party which consent shall not unreasonably be withheld.

(2) Nothing in this Contract confers or purports to confer on any third party any benefit or any right pursuant to the Contracts (Rights of Third Parties) Act 1999 to enforce any term of the Contract.

Sub-contracting 4

(1) The Contractor shall not sub-contract the whole of the Works without the prior written consent of the Employer.

(2) Except where otherwise provided in the Appendix to the Form of Tender the Contractor may sub-contract any part of the Works or their design. The extent of the work to be sub-contracted and the name and address of the sub-contractor must be notified in writing to the Engineer as soon as practicable and in any event not later than 14 days prior to the sub-contractor's entry on to the Site or in the case of design on appointment.

Provided that if not later than 7 days after receipt of such notification the Engineer for good reason objects to the Contractor in writing the sub-contractor so notified shall not be employed on or in connection with the Works. Such objection must be accompanied by reasons in writing.

(3) The employment of labour-only sub-contractors does not require notification to the Engineer under sub-clause (2) of this Clause.

(4) The Contractor shall be and remain liable under the Contract for all work sub-contracted by him and for acts defaults or neglects of any sub-contractor his agents servants or workpeople.

(5) The Engineer shall be at liberty after due warning in writing to require the Contractor to remove from the Works or their design any sub-contractor who misconducts himself or is incompetent or negligent in the performance of his duties or fails to conform with any particular provisions with regard to safety which may be set out in the Contract or persists in any conduct which is prejudicial to safety or health and such sub-contractor shall not be again employed upon the Works without the permission of the Engineer.

CONTRACT DOCUMENTS

Documents mutually explanatory

5 The several documents forming the Contract are to be taken as mutually explanatory of one another and in case of ambiguities or discrepancies the same shall be explained and adjusted by the Engineer who shall thereupon issue to the Contractor appropriate instructions in writing which shall be regarded as instructions issued in accordance with Clause 13.

Supply of documents

6 (1) Upon award of the Contract the following shall be furnished to the Contractor free of charge

(a) four copies of the Conditions of Contract Specification and (unpriced) bill of quantities and

(b) the number and type of copies as entered in the Appendix to the Form of Tender of all Drawings listed in the Specification.

(2) Upon acceptance by the Engineer in accordance with Clause 7(6) the Contractor shall supply to the Engineer four copies of all Drawings Specifications and other documents submitted by the Contractor. In addition the Contractor shall supply at the Employer's expense such further copies of such Drawings Specifications and other documents as the Engineer may request in writing for his use.

(3) Copyright of all Drawings Specifications and the Bill of Quantities (except the pricing thereof) supplied by the Employer or the Engineer shall not pass to the Contractor but the Contractor may obtain or make at his own expense any further copies required by him for the purposes of the Contract. Similarly copyright in all documents supplied by the Contractor under Clause 7(6) shall remain with the Contractor but the Employer and the Engineer shall have full power to reproduce and use the same for the purpose of completing operating maintaining and adjusting the Works.

Further Drawings Specifications and instructions

7 (1) The Engineer shall from time to time during the progress of the Works supply to the Contractor such modified or further Drawings Specifications and instructions as shall in the Engineer's opinion be necessary for the purpose of the proper and adequate construction and completion of the Works and the Contractor shall carry out and be bound by the same.

If such Drawings Specifications or instructions require any variation to any part of the Works the same shall be deemed to have been issued pursuant to Clause 51.

Contractor to provide further documents

(2) Where sub-clause (6) of this Clause applies the Engineer may require the Contractor to supply such further documents as shall in the Engineer's opinion be necessary for the purpose of the proper and adequate construction completion and maintenance of the Works and when accepted by the Engineer the Contractor shall be bound by the same.

Notice by Contractor

(3) The Contractor shall give adequate notice in writing to the Engineer of any further Drawing Specification or instruction that the Contractor may require for the construction and completion of the Works or otherwise under the Contract.

Delay in issue

(4) (a) If by reason of any failure or inability of the Engineer to issue at a time reasonable in all the circumstances Drawings Specifications or instructions requested by the Contractor and considered necessary by the Engineer in accordance with sub-clause (1) of this Clause the Contractor suffers delay or incurs additional cost then the Engineer shall take such delay into account in determining any extension of time to which the Contractor is entitled under Clause 44 and the Contractor shall subject to Clause 53 be paid in accordance with Clause 60 the amount of such cost as may be reasonable.

(b) If the failure of the Engineer to issue any Drawing Specification or instruction is caused in whole or in part by the failure of the Contractor after due notice in writing to submit drawings specifications or other documents which he is required to submit under the Contract the Engineer shall take into account such failure by the Contractor in taking any action under sub-clause (4)(a) of this Clause.

One copy of documents to be kept on Site

(5) One copy of the Drawings and Specification furnished to the Contractor as aforesaid and of all Drawings Specifications and other documents required to be provided by the Contractor under sub-clause (6) of this Clause shall at all reasonable times be available on the Site for inspection and use by the Engineer and the Engineer's Representative and by any other person authorized by the Engineer in writing.

Permanent Works designed by Contractor

(6) Where the Contract expressly provides that part of the Permanent Works shall be designed by the Contractor he shall submit to the Engineer for acceptance

(a) such drawings specifications calculations and other information as shall be necessary to satisfy the Engineer that the Contractor's design generally complies with the requirements of the Contract and

(b) operation and maintenance manuals together with as completed drawings of that part of the Permanent Works in sufficient detail to enable the Employer to operate maintain dismantle reassemble and adjust the Permanent Works incorporating that design. No certificate under Clause 48 covering any part of the Permanent Works designed by the Contractor shall be issued until manuals and drawings in such detail have been submitted to and accepted by the Engineer.

Responsibility unaffected by acceptance

(7) Acceptance by the Engineer in accordance with sub-clause (6) of this Clause shall not relieve the Contractor of any of his responsibilities under the Contract. The Engineer shall be responsible for the integration and co-ordination of the Contractor's design with the rest of the Works.

GENERAL OBLIGATIONS

Contractor's general responsibilities

8 (1) The Contractor shall subject to the provisions of the Contract

(a) construct and complete the Works and

(b) provide all labour materials Contractor's Equipment Temporary Works transport to and from and in or about the Site and everything whether of a temporary or permanent nature required in and for such construction and completion so far as the necessity for providing the same is specified in or reasonably to be inferred from the Contract.

Design responsibility

(2) The Contractor shall not be responsible for the design or specification of the Permanent Works or any part thereof (except as may be expressly provided in the Contract) or of any Temporary Works design supplied by the Engineer. The Contractor shall exercise all reasonable skill care and diligence in designing any part of the Permanent Works for which he is responsible.

Contractor responsible for safety of site operations

(3) The Contractor shall take full responsibility for the adequacy stability and safety of all site operations and methods of construction.

Form of Agreement **9**

The Contractor shall if called upon so to do enter into and execute an agreement to be prepared at the cost of the Employer in the form annexed to these Conditions.

Performance security **10**

(1) If the Contract requires the Contractor to provide security for the proper performance of the Contract he shall obtain and provide to the Employer such security in a sum not exceeding 10% of the Tender Total within 28 days of the award of the Contract. The security shall be provided by a body approved by the Employer and be in the Form of Bond annexed to these Conditions. The Contractor shall pay the cost of the security unless the Contract provides otherwise.

Dispute resolution upon security

(2) For the purposes of the dispute resolution provisions in such security

(a) the Employer shall be deemed to be a party to the security for the purpose of doing everything necessary to give effect to such provisions and

(b) any agreement decision award or other determination touching or concerning the relevant date for the discharge of the security shall be wholly without prejudice to the resolution or determination of any dispute between the Employer and the Contractor under Clause 66.

Provision and interpretation of information **11**

(1) As between the Employer and the Contractor and without prejudice to sub-clause (2) of this Clause information on

(a) the nature of the ground and subsoil and hydrological conditions and

(b) pipes and cables in on or over the ground

obtained by or on behalf of the Employer from investigations undertaken relevant to the Works shall only be taken into account to the extent that it was made available to the Contractor before the submission of his tender.

The Contractor shall be responsible for the interpretation of all such information for the purposes of constructing the Works and for any design which is the Contractor's responsibility under the Contract.

Inspection of Site

(2) The Contractor shall be deemed to have inspected and examined the Site and its surroundings and information available in connection therewith and to have satisfied himself so far as is practicable and reasonable before submitting his tender as to

(a) the form and nature thereof including the ground and sub-soil and hydrological conditions and

(b) the extent and nature of work and materials necessary for constructing and completing the Works and

(c) the means of communication with and access to the Site and the accommodation he may require

and in general to have obtained for himself all necessary information as to risks contingencies and all other circumstances which may influence or affect his tender.

Basis and sufficiency of tender

(3) The Contractor shall be deemed to have

(a) based his tender on his own inspection and examination as aforesaid and on all information whether obtainable by him or made available by the Employer and

(b) satisfied himself before submitting his tender as to the correctness and sufficiency of the rates and prices stated by him in the Bill of Quantities which shall (unless otherwise provided in the Contract) cover all his obligations under the Contract.

Adverse physical conditions and artificial obstructions

12 (1) If during the carrying out of the Works the Contractor encounters physical conditions (other than weather conditions or conditions due to weather conditions) or artificial obstructions which conditions or obstructions could not in his opinion reasonably have been foreseen by an experienced contractor the Contractor shall as early as practicable give written notice thereof to the Engineer.

Intention to claim

(2) If in addition the Contractor intends to make any claim for additional payment or extension of time arising from any such condition or obstruction he shall at the same time or as soon thereafter as may be reasonable inform the Engineer in writing pursuant to Clause 53 and/or Clause 44(1) as may be appropriate specifying the condition or obstruction to which the claim relates.

Measures being taken

(3) When giving notification or information in accordance with sub-clauses (1) and/or (2) of this Clause or as soon as practicable thereafter the Contractor shall give details of any anticipated effects of the condition or obstruction the measures he has taken is taking or is proposing to take their estimated cost and the extent of the anticipated delay in or interference with the carrying out of the Works.

Action by Engineer

(4) Following receipt of any notification under sub-clauses (1) or (2) or receipt of details in accordance with sub-clause (3) of this Clause the Engineer may if he thinks fit among other things

(a) require the Contractor to investigate and report upon the practicality cost and timing of alternative measures which may be available

(b) give written consent to measures notified under sub-clause (3) of this Clause with or without modification

(c) give written instructions as to how the physical conditions or artificial obstructions are to be dealt with

(d) order a suspension under Clause 40 or a variation under Clause 51.

Conditions reasonably foreseeable

(5) If the Engineer shall decide that the physical conditions or artificial obstructions could in whole or in part have been reasonably foreseen by an experienced contractor he shall so inform the Contractor in writing as soon as he shall have reached that decision but the value of any variation previously ordered by him pursuant to sub-clause (4)(d) of this Clause shall be ascertained in accordance with Clause 52 and included in the Contract Price.

Delay and extra cost

(6) Where an extension of time or additional payment is claimed pursuant to sub-clauses (1) and (2) of this Clause the Engineer shall if in his opinion such conditions or obstructions could not reasonably have been foreseen by an experienced contractor

(a) determine any delay which the Contractor has suffered and

(b) determine the amount of any costs which may reasonably have been incurred by the Contractor (together with a reasonable percentage addition thereto in respect of profit)

by reason of such conditions or obstructions and shall notify the Contractor accordingly with a copy to the Employer.

Any delay so determined shall forthwith be considered under Clause 44(3) for an appropriate extension of time and the Contractor shall subject to Clause 53 be paid in accordance with Clause 60 the amount so determined.

Work to be to satisfaction of Engineer **13**

(1) Save insofar as it is legally or physically impossible the Contractor shall construct and complete the Works in strict accordance with the Contract to the satisfaction of the Engineer and shall comply with and adhere strictly to the Engineer's instructions on any matter connected therewith (whether mentioned in the Contract or not). The Contractor shall take instructions only from the Engineer or subject to Clause 2(4) from his duly appointed delegate.

Mode and manner of construction

(2) The whole of the materials Contractor's Equipment and labour to be provided by the Contractor under Clause 8 and the mode manner and speed of construction of the Works are to be of a kind and conducted in a manner acceptable to the Engineer.

Delay and extra cost

(3) If in pursuance of Clause 5 or sub-clause (1) of this Clause the Engineer shall issue instructions which involve the Contractor in delay or disrupt his arrangements or methods of construction so as to cause him to incur cost beyond that reasonably to have been foreseen by an experienced contractor at the time of tender then the Engineer shall take such delay into account in determining any extension of time to which the Contractor is entitled under Clause 44 and the Contractor shall subject to Clause 53 be paid in accordance with Clause 60 the amount of such cost as may be reasonable except to the extent that such delay and extra cost result from the Contractor's default. Profit shall be added thereto in respect of any additional permanent or temporary work. If such instructions require any variation to any part of the Works the same shall be deemed to have been given pursuant to Clause 51.

Programme to be furnished

14 (1) (a) Within 21 days after the award of the Contract the Contractor shall submit to the Engineer for his acceptance a programme showing the order in which he proposes to carry out the Works having regard to the provisions of Clause 42(1).

(b) At the same time the Contractor shall also provide in writing for the information of the Engineer a general description of the arrangements and methods of construction which the Contractor proposes to adopt for the carrying out of the Works.

(c) Should the Engineer reject any programme under sub-clause (2)(b) of this Clause the Contractor shall within 21 days of such rejection submit a revised programme.

Action by Engineer

(2) The Engineer shall within 21 days after receipt of the Contractor's programme

(a) accept the programme in writing or

(b) reject the programme in writing with reasons or

(c) request the Contractor to supply further information to clarify or substantiate the programme or to satisfy the Engineer as to its reasonableness having regard to the Contractor's obligations under the Contract.

Provided that if none of the above actions is taken within the said period of 21 days the Engineer shall be deemed to have accepted the programme as submitted.

Provision of further information

(3) The Contractor shall within 21 days after receiving from the Engineer any request under sub-clause (2)(c) of this Clause or within such further period as the Engineer may allow provide the further information requested failing which the relevant programme shall be deemed to be rejected.

Upon receipt of such further information the Engineer shall within a further 21 days accept or reject the programme in accordance with sub-clauses (2)(a) or (2)(b) of this Clause.

Revision of programme

(4) Should it appear to the Engineer at any time that the actual progress of the work does not conform with the accepted programme referred to in sub-clause (1) of this Clause the Engineer shall be entitled to require the Contractor to produce a revised programme showing such modifications to the original programme as may be necessary to ensure completion of the Works or any Section within the time for completion as defined in Clause 43 or extended time granted pursuant to Clause 44. In such event the Contractor shall submit his revised programme within 21 days or within such further period as the Engineer may allow. Thereafter the provisions of sub-clauses (2) and (3) of this Clause shall apply.

Design criteria

(5) The Engineer shall provide to the Contractor such design criteria relevant to the Permanent Works or any Temporary Works design supplied by the Engineer as may be necessary to enable the Contractor to comply with sub-clauses (6) and (7) of this Clause.

Methods of construction

(6) If requested by the Engineer the Contractor shall submit at such times and in such further detail as the Engineer may reasonably require information pertaining to the methods of construction (including Temporary Works and the use of Contractor's Equipment) which the Contractor proposes to adopt or use and calculations of stresses strains and deflections that will arise in the Permanent Works or any parts thereof during construction so as to enable the Engineer to decide whether if these methods are adhered to the Works can be constructed and completed in accordance with the Contract and without detriment to the Permanent Works when completed.

Engineer's consent

(7) The Engineer shall inform the Contractor in writing within 21 days after receipt of the information submitted in accordance with sub-clauses (1)(b) and (6) of this Clause either

(a) that the Contractor's proposed methods have the consent of the Engineer or

(b) in what respects in the opinion of the Engineer they fail to meet the requirements of the Contract or will be detrimental to the Permanent Works.

In the latter event the Contractor shall take such steps or make such changes in the said methods as may be necessary to meet the Engineer's requirements and to obtain his consent. The Contractor shall not change the methods which have received the Engineer's consent without the further consent in writing of the Engineer which shall not be unreasonably withheld.

Delay and extra cost

(8) If the Contractor unavoidably incurs delay or extra cost because

(a) the Engineer's consent to the proposed methods of construction is unreasonably delayed or

(b) the Engineer's requirements pursuant to sub-clause (7) of this Clause or any limitations imposed by any of the design criteria supplied by the Engineer pursuant to sub-clause (5) of this Clause could not reasonably have been foreseen by an experienced contractor at the time of tender

then the Engineer shall take such delay into account in determining any extension of time to which the Contractor is entitled under Clause 44 and the Contractor shall subject to Clause 53 be paid in accordance with Clause 60 the amount of such cost as may be reasonable except to the extent that such delay and extra cost result from the Contractor's default. Profit shall be added thereto in respect of any additional permanent or temporary work.

Responsibility unaffected by acceptance or consent

(9) Acceptance (or deemed acceptance) by the Engineer of the Contractor's programme in accordance with sub-clauses (2)(3) or (4) of this Clause and the consent of the Engineer to the Contractor's proposed methods of construction in accordance with sub-clause (7) of this Clause shall not relieve the Contractor of any of his duties or responsibilities under the Contract.

Contractor's superintendence **15**

(1) The Contractor shall provide all necessary superintendence during the construction and completion of the Works and for as long thereafter as the Engineer may reasonably consider necessary.

Such superintendence shall be given by sufficient persons having adequate knowledge of the operations to be carried out (including the methods and techniques required the hazards likely to be encountered and methods of preventing accidents) for the satisfactory and safe construction of the Works.

Contractor's agent

(2) The Contractor or a competent and authorized agent or representative approved of in writing by the Engineer (which approval may at any time be withdrawn) is to be constantly on the Works and shall give his whole time to the superintendence of the same. Such authorized agent or representative shall be in full charge of the Works and shall receive on behalf of the Contractor directions and instructions from the Engineer or (subject to the limitations of Clause 2) the Engineer's Representative. The Contractor or such authorized agent or representative shall be responsible for the safety of all operations.

Removal of Contractor's employees **16**

The Contractor shall employ or cause to be employed in and about the construction and completion of the Works and in the superintendence thereof only persons who are careful skilled and experienced in their several trades and callings.

The Engineer shall be at liberty to object to and require the Contractor to remove or cause to be removed from the Works any person employed thereon who in the opinion of the Engineer mis-conducts himself or is incompetent or negligent in the performance of his duties or fails to conform with any particular provisions with regard to safety which may be set out in the Contract or persists in any conduct which is prejudicial to safety or health and such persons shall not be again employed upon the Works without the permission of the Engineer.

Setting-out **17**

(1) The Contractor shall be responsible for the true and proper setting-out of the Works and for the correctness of the position levels dimensions and alignment of all parts of the Works and for the provision of all necessary instruments appliances and labour in connection therewith.

(2) If at any time during the progress of the Works any error shall appear or arise in the position levels dimensions or alignment of any part of the Works the Contractor on being required so to do by the Engineer shall at his own cost rectify such error to the satisfaction of the Engineer unless such error is based on incorrect data supplied in writing by the Engineer or the Engineer's Representative in which case the cost of rectifying the same shall be borne by the Employer.

(3) The checking of any setting-out or of any line or level by the Engineer or the Engineer's Representative shall not in any way relieve the Contractor of his responsibility for the correctness thereof and the Contractor shall carefully protect and preserve all bench-marks sight rails pegs and other things used in setting out the Works.

Boreholes and exploratory excavation

18 If at any time during the construction of the Works the Engineer shall require the Contractor to make boreholes or to carry out exploratory excavation such requirement shall be ordered in writing and shall be deemed to be a variation under Clause 51 unless a Provisional Sum or Prime Cost Item in respect of such anticipated work shall have been included in the Bill of Quantities.

Safety and security

19 (1) The Contractor shall throughout the progress of the Works have full regard for the safety of all persons entitled to be upon the Site and shall keep the Site (so far as the same is under his control) and the Works (so far as the same are not completed or occupied by the Employer) in an orderly state appropriate to the avoidance of danger to such persons.

Lighting fencing and watching

(2) The Contractor shall among other things in connection with the Works provide and maintain at his own cost all lights guards fencing warning signs and watching when and where necessary or required by the Engineer or the Engineer's Representative or by any statutory or other competent authority for the protection of the Works or for the safety and convenience of the public or others.

Employer's responsibilities

(3) If the Employer carries out work on the Site with his own workpeople he shall in respect of such work

(a) have full regard for the safety of all persons entitled to be upon the Site and

(b) keep the Site in an orderly state appropriate to the avoidance of danger to such persons.

If the Employer employs other contractors on the Site he shall require them to have the same regard for safety and avoidance of danger.

Care of the Works

20 (1) (a) The Contractor shall save as in paragraph (b) hereof and subject to sub-clause (2) of this Clause takes responsibility for the care of the Works and materials plant and equipment for incorporation therein from the Works Commencement Date until the date of issue of a Certificate of Substantial Completion for the whole of the Works when the responsibility for the said care shall pass to the Employer.

(b) If the Engineer issues a Certificate of Substantial Completion for any Section or part of the Permanent Works the Contractor shall cease to be responsible for the care of that Section or part from the date of issue of that Certificate of Substantial Completion when the responsibility for the care of that Section or part shall pass to the Employer.

(c) The Contractor shall take full responsibility for the care of any work and materials plant and equipment for incorporation therein which he undertakes during the Defects Correction Period until such work has been completed.

Excepted Risks

(2) The Excepted Risks for which the Contractor is not liable are loss or damage to the extent that it is due to

(a) the use or occupation by the Employer his agents servants or other contractors (not being employed by the Contractor) of any part of the Permanent Works

(b) any fault defect error or omission in the design of the Works (other than a design provided by the Contractor pursuant to his obligations under the Contract)

(c) riot war invasion act of foreign enemies or hostilities (whether war be declared or not)

(d) civil war rebellion revolution insurrection or military or usurped power

(e) ionizing radiations or contamination by radioactivity from any nuclear fuel or from any nuclear waste from the combustion of nuclear fuel radioactive toxic explosive or other hazardous properties of any explosive nuclear assembly or nuclear component thereof and

(f) pressure waves caused by aircraft or other aerial devices travelling at sonic or supersonic speeds.

Rectification of loss or damage

(3) (a) In the event of any loss or damage to

(i) the Works or any Section or part thereof or

(ii) materials plant or equipment for incorporation therein

while the Contractor is responsible for the care thereof (except as provided in sub-clause (2) of this Clause) the Contractor shall at his own cost rectify such loss or damage so that the Permanent Works conform in every respect with the provisions of the Contract and the Engineer's instructions. The Contractor shall also be liable for any loss or damage to the Works occasioned by him in the course of any operations carried out by him for the purpose of complying with his obligations under Clauses 49 and 50.

(b) Should any such loss or damage arise from any of the Excepted Risks defined in sub-clause (2) of this Clause the Contractor shall if and to the extent required by the Engineer rectify the loss or damage at the expense of the Employer.

(c) In the event of loss or damage arising from an Excepted Risk and a risk for which the Contractor is responsible under sub-clause (1)(a) of this Clause then the cost of rectification shall be apportioned accordingly.

Insurance of Works etc. **21**

(1) The Contractor shall without limiting his or the Employer's obligations and responsibilities under Clause 20 insure in the joint names of the Contractor and the Employer the Works together with materials plant and equipment for incorporation therein to the full replacement cost plus an additional 10% to cover any additional costs that may arise incidental to the rectification of any loss or damage including professional fees cost of demolition and removal of debris.

Extent of cover

(2) (a) The insurance required under sub-clause (1) of this Clause shall cover the Employer and the Contractor against all loss or damage from whatsoever cause arising other than the Excepted Risks defined in Clause 20(2) from the Works Commencement Date until the date of issue of the relevant Certificate of Substantial Completion.

(b) The insurance shall extend to cover any loss or damage arising during the Defects Correction Period from a cause occurring prior to the issue of any Certificate of Substantial Completion and any loss or damage occasioned by the Contractor in the course of any operation carried out by him for the purpose of complying with his obligations under Clauses 49, 50 and 51.

(c) Nothing in this Clause shall render the Contractor liable to insure against the necessity for the repair or reconstruction of any work constructed with materials or workmanship not in accordance with the requirements of the Contract.

(d) Any amounts not insured or not recovered from insurers whether as excesses carried under the policy or otherwise shall be borne by the Contractor or the Employer in accordance with their respective responsibilities under Clause 20.

Damage to persons and property

22 (1) The Contractor shall except if and so far as the Contract provides otherwise and subject to the exceptions set out in sub-clause (2) of this Clause indemnify and keep indemnified the Employer against all losses and claims in respect of

(a) death of or injury to any person or

(b) loss of or damage to any property (other than the Works)

which may arise out of or in consequence of the construction of the Works and the remedying of any defects therein and against all claims demands proceedings damages costs charges and expenses whatsoever in respect thereof or in relation thereto.

Exceptions

(2) The exceptions referred to in sub-clause (1) of this Clause which are the responsibility of the Employer are

(a) damage to crops being on the Site (save in so far as possession has not been given to the Contractor)

(b) the use or occupation of land provided by the Employer for the purposes of the Contract (including consequent losses of crops) or interference whether temporary or permanent with any right of way light air or water or other easement or quasi-easement which are the unavoidable result of the construction of the Works in accordance with the Contract

(c) the right of the Employer to construct the Works or any part thereof on over under in or through any land

(d) damage which is the unavoidable result of the construction of the Works in accordance with the Contract and

(e) death of or injury to persons or loss of or damage to property resulting from any act neglect or breach of statutory duty done or committed by the Employer his agents servants or other contractors (not being employed by the Contractor) or for or in respect of any claims demands proceedings damages costs charges and expenses in respect thereof or in relation thereto.

Indemnity by Employer

(3) The Employer shall subject to sub-clause (4) of this Clause indemnify the Contractor against all claims demands proceedings damages costs charges and expenses in respect of the matters referred to in the exceptions defined in sub-clause (2) of this Clause.

Shared responsibility

(4) (a) The Contractor's liability to indemnify the Employer under sub-clause (1) of this Clause shall be reduced in proportion to the extent that the act or neglect of the Employer his agents servants or other contractors (not being employed by the Contractor) may have contributed to the said death injury loss or damage.

(b) The Employer's liability to indemnify the Contractor under sub-clause (3) of this Clause in respect of matters referred to in sub-clause (2)(e) of this Clause shall be reduced in proportion to the extent that the act or neglect of the Contractor or his sub-contractors servants or agents may have contributed to the said death injury loss or damage.

Third party insurance

23 (1) The Contractor shall without limiting his or the Employer's obligations and responsibilities under Clause 22 insure in the joint names of the Contractor and the Employer against liabilities for death of or injury to any person (other than any operative or other person in the employment of the Contractor or any of his sub-contractors) or loss of or damage to any property (other than the Works) arising out of the performance of the Contract other than those liabilities arising out of the exceptions defined in Clause 22(2)(a) (b) (c) and (d).

Cross liability clause

(2) The insurance policy shall include a cross liability clause such that the insurance shall apply to the Contractor and to the Employer as separate insured.

Amount of insurance

(3) Such insurance shall be for at least the amount stated in the Appendix to the Form of Tender.

Accident or injury to operatives etc.

24 The Employer shall not be liable for or in respect of any damages or compensation payable at law in respect or in consequence of any accident or injury to any operative or other person in the employment of the Contractor or any of his sub-contractors save and except to the extent that such accident or injury results from or is contributed to by any act or default of the Employer his agents or servants and the Contractor shall indemnify and keep indemnified the Employer against all such damages and compensation (save and except as aforesaid) and against all claims demands proceedings costs charges and expenses whatsoever in respect thereof or in relation thereto.

Evidence and terms of insurance

25 (1) The Contractor shall provide satisfactory evidence to the Employer prior to the Works Commencement Date that the insurances required under the Contract have been effected and shall if so required produce the insurance policies for inspection. The terms of all such insurances shall be subject to the approval of the Employer (which approval shall not unreasonably be withheld). The Contractor shall upon request produce to the Employer receipts for the payment of current insurance premiums.

Excesses

(2) Any excesses on the policies of insurance effected under Clauses 21 and 23 shall be no greater than those stated in the Appendix to the Form of Tender.

Remedy on Contractor's failure to insure

(3) If the Contractor shall fail upon request to produce to the Employer satisfactory evidence that there is in force any of the insurances required under the Contract then the Employer may effect and keep in force any such insurance and pay such premium or premiums as may be necessary for that purpose and from time to time deduct the amount so paid from any monies due or which may become due to the Contractor or recover the same as a debt due from the Contractor.

Compliance with policy conditions

(4) Both the Employer and the Contractor shall comply with all conditions laid down in the insurance policies. Should the Contractor or the Employer fail to comply with any condition imposed by the insurance policies effected pursuant to the Contract each shall indemnify the other against all losses and claims arising from such failure.

Giving of notices and payment of fees

26 (1) Except where otherwise provided in the Contract the Contractor shall give all notices and pay all fees required to be given or paid by any Act of Parliament or any Regulation or Bye-law of any local or other statutory authority in relation to the construction and completion of the Works and by the rules and regulations of all public bodies and companies whose property or rights are or may be affected in any way by the Works.

Repayment by Employer

(2) The Employer shall repay or allow to the Contractor all such sums as the Engineer shall certify to have been properly payable and paid by the Contractor in respect of such fees and also all rates and taxes paid by the Contractor in respect of the Site or any part thereof or anything constructed or erected thereon or on any part thereof or any temporary structures situated elsewhere but used exclusively for the purposes of the Works or any structures used temporarily and exclusively for the purposes of the Works.

Contractor to conform with Statutes etc.

(3) The Contractor shall ascertain and conform in all respects with the provisions of any general or local Act of Parliament and the Regulations and Bye-laws of any local or other statutory authority which may be applicable to the Works and with such rules and regulations of public bodies and companies as aforesaid and shall keep the Employer indemnified against all penalties and liability of every kind for breach of any such Act Regulation or Bye-law. Provided always that

(a) the Contractor shall not be required to indemnify the Employer against the consequences of any such breach which is the unavoidable result of complying with the Contract or instructions of the Engineer

(b) if the Contract or instructions of the Engineer shall at any time be found not to be in conformity with any such Act Regulation or Bye-law the Engineer shall issue such instructions including the ordering of a variation under Clause 51 as may be necessary to ensure conformity with such Act Regulation or Bye-law and

(c) the Contractor shall not be responsible for obtaining any planning permission which may be necessary in respect of the Permanent Works in their final position or of any Temporary Works designed by the Engineer in their designated position on Site. The Employer hereby warrants that all such permissions have been or will in due time be obtained.

(4) If the Contractor incurs delay or extra cost arising from matters referred to in sub-clause (3)(b) or failure of the Employer to comply with sub-clause (3)(c) of this Clause the Engineer shall take such delay into account in determining any extension of time to which the Contractor may be entitled under Clause 44 and the Contractor shall subject to Clause 53 be paid in accordance with Clause 60 the amount of such extra cost as may be reasonable except to the extent that such delay or extra cost result from the Contractor's default.

New Roads and Street Works Act 1991—Definitions

27 (1) (a) In this Clause "the Act" shall mean the New Roads and Street Works Act 1991 and any statutory modification or re-enactment thereof for the time being in force.

(b) For the purpose of obtaining any license under the Act required for the Permanent Works the undertaker shall be the Employer who for the purposes of the Act will be the licensee.

(c) All other expressions common to the Act and to this Clause shall have the same meaning as those assigned to them by the Act.

Licences

(2) (a) The Employer shall obtain any street works licence and any other consent licence or permission that may be required for the carrying out of the Permanent Works and shall supply the Contractor with copies thereof including details of any conditions or limitations imposed.

Provided that where the license contains a prohibition against assignment which is notified to the Contractor then the Contractor shall give the Employer all notices required to be given by the undertaker and shall indemnify the Employer from and against all costs and charges which may arise from any failure by him so to do.

(b) Any condition or limitation including any requirement as to notice in any licence obtained after the award of the Contract shall be deemed to be an instruction under Clause 13.

Notices

(3) The Contractor shall be responsible for giving to any relevant authority any required notice (or advance notice where prescribed) of his proposal to commence any work. A copy of each such notice shall be given to the Employer.

Patent rights

28 (1) The Contractor shall indemnify and keep indemnified the Employer from and against all claims and proceedings for or on account of infringement of any patent right design trademark or name or other protected right in respect of any

(a) Contractor's Equipment used for or in connection with the Works

(b) materials plant and equipment for incorporation in the Works

and from and against all claims demands proceedings damages costs charges and expenses whatsoever in respect thereof or in relation thereto except where such infringement results from compliance with the design or Specification provided other than by the Contractor. In the latter event the Employer shall indemnify the Contractor from and against all claims and proceedings for or on account of infringement of any patent right design trademark or name or other protected right aforesaid.

Royalties

(2) Except where otherwise stated the Contractor shall pay all tonnage and other royalties rent and other payments or compensation (if any) for getting stone sand gravel clay or other materials required for the Works.

Interference with traffic and adjoining properties

29 (1) All operations necessary for the construction and completion of the Works shall so far as compliance with the requirements of the Contract permits be carried on so as not to interfere unnecessarily or improperly with

(a) the convenience of the public or

(b) the access to public or private roads footpaths or properties whether in the possession of the Employer or of any other person and with the use or occupation thereof.

The Contractor shall indemnify and keep indemnified the Employer in respect of all claims demands proceedings damages costs charges and expenses whatsoever arising out of or in relation to any such matters.

Noise disturbance and pollution

(2) All work shall be carried out without unreasonable noise disturbance or other pollution.

Indemnity by Contractor

(3) To the extent that noise disturbance or other pollution is not the unavoidable consequence of constructing and completing the Works or performing the Contract the Contractor shall indemnify the Employer from and against any liability for damages on that account and against all claims demands proceedings damages costs charges and expenses whatsoever in regard or in relation to such liability.

Indemnity by Employer

(4) The Employer shall indemnify the Contractor from and against any liability for damages on account of noise disturbance or other pollution which is the unavoidable consequence of carrying out the Works and from and against all claims demands proceedings damages costs charges and expenses whatsoever in regard or in relation to such liability.

Avoidance of damage to highways etc.

30 (1) The Contractor shall use every reasonable means to prevent any of the highways or bridges communicating with or on the routes to the Site from being subjected to extraordinary traffic within the meaning of the Highways Act 1980 or in Scotland the Roads (Scotland) Act 1984 or any statutory modification or re-enactment thereof by any traffic of the Contractor or any of his sub-contractors.

In particular shall select routes and use vehicles and restrict and distribute loads so that any such extraordinary traffic as will inevitably arise from the moving of Contractor's Equipment and materials or manufactured or fabricated articles from and to the Site shall be limited as far as reasonably possible and so that no unnecessary damage or injury may be occasioned to such highways and bridges.

Transport of Contractor's Equipment

(2) Save insofar as the Contract otherwise provides the Contractor shall be responsible for and shall pay the cost of strengthening any bridges or altering or improving any highway communicating with the Site to facilitate the movement of Contractor's Equipment or Temporary Works required in the carrying out of the Works.

The Contractor shall indemnify and keep indemnified the Employer against all claims for damage to any highway or bridge communicating with the Site caused by such movement including such claims as may be made by any competent authority directly against the Employer pursuant to any Act of Parliament or other Statutory Instrument and shall negotiate and pay all claims arising solely out of such damage.

Transport of materials

(3) (a) If notwithstanding sub-clause (1) of this Clause any damage occurs to any bridge or highway communicating with the Site arising from the transport of materials or manufactured or fabricated articles being or

intended to form part of the Permanent Works or any Temporary Works designed by the Engineer the Contractor shall notify the Engineer as soon as he becomes aware of such damage or as soon as he receives any claim from the authority entitled to make such claim.

(b) Where under any Act of Parliament or other Statutory Instrument the haulier of such materials or manufactured or fabricated articles is required to indemnify the highway authority against damage the Employer shall not be liable for any costs charges or expenses in respect thereof or in relation thereto.

(c) In other cases the Employer shall negotiate the settlement of and pay all sums due in respect of such claim and shall indemnify the Contractor in respect thereof and in respect of all claims demands proceedings damages costs charges and expenses in relation thereto.

Provided always that if and so far as any such claim or part thereof is in the opinion of the Engineer due to any failure on the part of the Contractor to observe and perform his obligations under sub-clause (1) of this Clause then the amount certified by the Engineer to be due to such failure shall be paid by the Contractor to the Employer or deducted from any sum due or which may become due to the Contractor.

Facilities for other contractors

31 (1) The Contractor shall in accordance with the requirements of the Engineer or Engineer's Representative afford all reasonable facilities for any other contractors employed by the Employer and their workmen and for the workmen of the Employer and of any other properly authorised authorities or statutory bodies who may be employed in the carrying out on or near the Site of any work not in the Contract or of any contract which the Employer may enter into in connection with or ancillary to the Works.

Delay and extra cost

(2) If compliance with sub-clause (1) of this Clause involves the Contractor in delay or cost beyond that reasonably to have been foreseen by an experienced contractor at the time of tender then the Engineer shall take such delay into account in determining any extension of time to which the Contractor is entitled under Clause 44 and the Contractor shall subject to Clause 53 be paid in accordance with Clause 60 the amount of such cost as may be reasonable. Profit shall be added thereto in respect of any additional permanent or temporary work.

Fossils etc.

32 All fossils coins articles of value or antiquity and structures or other remains or things of geological or archaeological interest discovered on the Site shall as between the Employer and the Contractor be deemed to be the absolute property of the Employer and the Contractor shall take reasonable precautions to prevent his workmen or any other persons from removing or damaging any such article or thing and shall immediately upon discovery thereof and before removal acquaint the Engineer of such discovery and carry out at the expense of the Employer the Engineer's orders as to the disposal of the same.

Clearance of Site on completion

33 On completion of the Works the Contractor shall clear away and remove from the Site all Contractor's Equipment surplus material rubbish and Temporary Works of every kind and leave the whole of the Site and Permanent Works clean and in a workmanlike condition to the satisfaction of the Engineer.

34 (Not used)

Returns of labour and Contractor's Equipment

35 The Contractor shall if required by the Engineer deliver to the Engineer or the Engineer's Representative a return in such form and at such intervals as the Engineer may prescribe showing in detail the numbers of the several classes of labour from time to time employed by the Contractor on the Site and such information respecting Contractor's Equipment as the Engineer may require. The Contractor shall require his sub-contractors to observe the provisions of this Clause.

MATERIALS AND WORKMANSHIP

Quality of materials and workmanship and tests

36 (1) All materials and workmanship shall be of the respective kinds described in the Contract and in accordance with the Engineer's instructions and shall be subjected from time to time to such tests as the Engineer may direct at the place of manufacture or fabrication or on the Site or such other place or places as may be specified in the Contract. The Contractor shall provide such assistance instruments machines labour and materials as are normally required for examining measuring and testing any work and the quality weight or quantity of any materials used and shall supply samples of materials before incorporation in the Works for testing as may be selected and required by the Engineer.

Cost of samples

(2) All samples shall be supplied by the Contractor at his own cost if the supply thereof is clearly intended by or provided for in the Contract but if not then at the cost of the Employer.

Cost of tests

(3) The cost of making any test shall be borne by the Contractor if such test is clearly intended by or provided for in the Contract and (in the cases only of a test under load or of a test to ascertain whether the design of any finished or partially finished work is appropriate for the purposes which it was intended to fulfil) is particularized in the Contract in sufficient detail to enable the Contractor to have priced or allowed for the same in his tender.

(4) In all other cases the cost of making any test shall be borne by the Employer unless the need for such test results from the Contractor's default or from failure on the part of the Contractor to observe and perform his obligations under the Contract.

Access to Site

37 The Engineer and any person authorized by him shall at all times have access to the Works and to the Site and to all workshops and places where work is being prepared or whence materials manufactured articles and machinery are being obtained for the Works and the Contractor shall afford every facility for and every assistance in obtaining such access or the right to such access.

Examination of work before covering up

38 (1) No work shall be covered up or put out of view without the consent of the Engineer and the Contractor shall afford full opportunity for the Engineer to examine and measure any work which is about to be covered up or put out of view and to examine foundations before permanent work is placed thereon. The Contractor shall give due notice to the Engineer whenever any such work or foundations is or are ready or about to be ready for examination and the Engineer shall without unreasonable delay unless he considers it unnecessary and advises the Contractor accordingly attend for the purpose of examining and measuring such work or of examining such foundations.

Uncovering and making openings

(2) The Contractor shall uncover any part or parts of the Works or make openings in or through the same as the Engineer may from time to time direct and shall reinstate and make good such part or parts to the satisfaction of the Engineer. If any such part or parts have been covered up or put out of view after compliance with the requirements of sub-clause (1) of this Clause and are found to have been carried out in accordance with the Contract the cost of uncovering making openings in or through reinstating and making good the same shall be borne by the Employer but in any other case all such cost shall be borne by the Contractor.

Removal of unsatisfactory work and materials

39 (1) The Engineer shall during the progress of the Works have power to instruct in writing

(a) the removal from the Site within such time or times specified in the instruction of any materials which in the opinion of the Engineer are not in accordance with the Contract and

(b) substitution with materials in accordance with the Contract and/or

(c) the removal and proper replacement (notwithstanding any previous test thereof or interim payment therefor) of any work which in respect of

(i) materials or workmanship or

(ii) design by the Contractor or for which he is responsible

is not in the opinion of the Engineer in accordance with the Contract.

Default of Contractor in compliance

(2) In case of default on the part of the Contractor in carrying out such instruction the Employer shall be entitled to employ and pay other persons to carry out the same and all costs consequent thereon or incidental thereto as determined by the Engineer shall be recoverable from the Contractor by the Employer and may be deducted by the Employer from any monies due or to become due to him and the Engineer shall notify the Contractor accordingly with a copy to the Employer.

Failure to disapprove

(3) Failure of the Engineer or any person acting under him pursuant to Clause 2 to disapprove any work or materials shall not prejudice the power of the Engineer or any such person subsequently to take action under this Clause.

Suspension of Work

40 (1) The Contractor shall on the written order of the Engineer suspend the progress of the Works or any part thereof for such time or times and in such manner as the Engineer may consider necessary and shall during such suspension properly protect and secure the work so far as is necessary in the opinion of the Engineer. Except to the extent that such suspension is

(a) otherwise provided for in the Contract or

(b) necessary by reason of weather conditions or by some default on the part of the Contractor or

(c) necessary for the proper construction and completion or for the safety of the Works or any part thereof in as much as such necessity does not arise from any act or default of the Engineer or the Employer or from any of the Excepted Risks defined in Clause 20(2) then if compliance with the Engineer's instructions under this clause involves the Contractor in delay or extra cost the Engineer shall take such delay into account in determining any extension of time to which the Contractor is entitled under Clause 44 and the Contractor shall subject to Clause 53 be paid in accordance with Clause 60 the amount of such extra cost as may be reasonable. Profit shall be added thereto.

Suspension lasting more than three months

(2) If the progress of the Works or any part thereof is suspended on the written order of the Engineer and if permission to resume work is not given by the Engineer within a period of 3 months from the date of suspension then the Contractor may unless such suspension is otherwise provided for in the Contract or continues to be necessary by reason of some default on the part of the Contractor serve a written notice on the Engineer requiring permission within 28 days from the receipt of such notice to proceed with the Works or that part thereof in regard to which progress is suspended. If within the said 28 days the Engineer does not grant such permission the Contractor by a further written notice so served may (but is not bound to) elect to treat the suspension where it affects part only of the Works as an omission of such part under Clause 51 or where it affects the whole Works as an abandonment of the Contract by the Employer.

COMMENCEMENT TIME AND DELAYS

Works Commencement Date

41 (1) The Works Commencement Date shall be

(a) the date specified in the Appendix to the Form of Tender or if no date is specified

(b) a date between 14 and 28 days of the award of the Contract to be notified to the Contractor by the Engineer in writing or

(c) such other date as may be agreed between the parties.

Start of Works

(2) The Contractor shall start the Works on or as soon as is reasonably practicable after the Works Commencement Date. Thereafter the Contractor shall proceed with the Works with due expedition and without delay in accordance with the Contract.

Possession of Site and access

42 (1) The Contract may prescribe

(a) the extent of portions of the Site of which the Contractor is to be given possession from time to time

(b) the order in which such portions of the Site shall be made available to the Contractor

(c) the availability and the nature of the access which is to be provided by the Employer

(d) the order in which the Works shall be constructed.

(2) (a) Subject to sub-clause (1) of this Clause the Employer shall give to the Contractor on the Works Commencement Date possession of the whole of the Site together with such access thereto as may be necessary to enable the Contractor to commence and proceed with the construction of the Works.

(b) Thereafter the Employer shall during the course of the Works give to the Contractor such further access in accordance with the Contract as is necessary to enable the Contractor to proceed with the construction of the Works with due despatch.

Failure to give possession

(3) If the Contractor suffers delay and/or incurs extra cost from failure on the part of the Employer to give possession or access in accordance with the terms of this Clause the Engineer shall take such delay into account in determining any extension of time to which the Contractor is entitled under Clause 44 and the Contractor shall subject to Clause 53 be paid in accordance with Clause 60 the amount of any extra cost to which he may be entitled. Profit shall be added thereto in respect of any additional permanent or temporary work.

Access and facilities provided by the Contractor

(4) The Contractor shall bear all costs and charges for any access required by him additional to those provided by the Employer. The Contractor shall also provide at his own cost any additional facilities outside the Site required by him for the purposes of the Works.

Time for completion

43 The whole of the Works and any Section required to be completed within a particular time as stated in the Appendix to the Form of Tender shall be substantially completed within the time so stated (or such extended time as may be allowed under Clause 44 or revised time agreed under Clause 46(3)) calculated from the Works Commencement Date.

Extension of time for completion

44 (1) Should the Contractor consider that

(a) any variation ordered under Clause 51(1) or

(b) increased quantities referred to in Clause 51(4) or

(c) any cause of delay referred to in these Conditions or

(d) exceptional adverse weather conditions or

(e) any delay impediment prevention or default by the Employer or

(f) other special circumstances of any kind whatsoever which may occur

be such as to entitle him to an extension of time for the substantial completion of the Works or any Section thereof he shall within 28 days after the cause of any delay has arisen or as soon thereafter as is reasonable deliver to the Engineer full and detailed particulars in justification of the period of extension claimed in order that the claim may be investigated at the time.

Assessment of delay

(2) (a) The Engineer shall upon receipt of such particulars consider all the circumstances known to him at that time and make an assessment of the delay (if any) that has been suffered by the Contractor as a result of the alleged cause and shall so notify the Contractor in writing.

(b) The Engineer may in the absence of any claim make an assessment of the delay that he considers has been suffered by the Contractor as a result of any of the circumstances listed in sub-clause (1) of this Clause and shall so notify the Contractor in writing.

Interim grant of extension of time

(3) Should the Engineer consider that the delay suffered fairly entitles the Contractor to an extension of the time for the substantial completion of the Works or any Section thereof such interim extension shall be granted forthwith and be notified to the Contractor in writing with a copy to the Employer. In the event that the Contractor has made a claim for an extension of time but the Engineer does not consider the Contractor entitled to an extension of time he shall so inform the Contractor without delay.

Assessment at due date for completion

(4) The Engineer shall not later than 14 days after the due date or extended date for completion of the Works or any Section thereof (and whether or not the Contractor shall have made any claim for an extension of time) consider all the circumstances known to him at that time and take action similar to that provided for in sub-clause (3) of this Clause. Should the Engineer consider that the Contractor is not entitled to an extension of time he shall so notify the Employer and the Contractor.

Final determination of extension

(5) The Engineer shall within 28 days of the issue of the Certificate of Substantial Completion for the Works or for any Section thereof review all the circumstances of the kind referred to in sub-clause (1) of this Clause and shall finally determine and certify to the Contractor with a copy to the Employer the overall extension of time (if any) to which he considers the Contractor entitled in respect of the Works or the relevant Section. No such final review of the circumstances shall result in a decrease in any extension of time already granted by the Engineer pursuant to sub-clauses (3) or (4) of this Clause.

Night and Sunday work **45**

Subject to any provision to the contrary contained in the Contract none of the Works shall be carried out during the night or on Sundays without the permission in writing of the Engineer save when the work is unavoidable or absolutely necessary for the saving of life or property or for the safety of the Works in which case the Contractor shall immediately advise the Engineer or the Engineer's Representative. Provided always that this Clause shall not be applicable in the case of any work which it is customary to carry out outside normal working hours or by rotary or double shifts.

Rate of progress **46**

(1) If for any reason which does not entitle the Contractor to an extension of time the rate of progress of the Works or any Section is at any time in the opinion of the Engineer too slow to ensure substantial completion by the time or extended time for completion prescribed by Clause 43 and 44 as appropriate or the revised time for completion agreed under sub-clause (3) of this Clause the Engineer shall notify the Contractor in writing and the Contractor shall thereupon take such steps as are necessary and to which the Engineer may consent to expedite the progress so as substantially to complete the Works or such Section by that prescribed time or extended time. The Contractor shall not be entitled to any additional payment for taking such steps.

Permission to work at night or on Sundays

(2) If as a result of any notice given by the Engineer under sub-clause (1) of this Clause the Contractor seeks the Engineer's permission to do any work on Site at night or on Sundays such permission shall not be unreasonably refused.

Provision for accelerated completion

(3) If the Contractor is requested by the Employer or the Engineer to complete the Works or any Section within a revised time being less than the time or extended time for completion prescribed by Clauses 43 and 44 as appropriate and the Contractor agrees so to do then any special terms and conditions of payment shall be agreed between the Contractor and the Employer before any such action is taken.

LIQUIDATED DAMAGES FOR DELAY

Liquidated damages for delay in substantial completion of the whole of the Works

47 (1) (a) Where the whole of the Works is not divided into Sections the Appendix to the Form of Tender shall include a sum which represents the Employer's genuine pre-estimate (expressed per week or per day as the case may be) of the damages likely to be suffered by him if the whole of the Works is not substantially completed within the time prescribed by Clause 43 or by any extension thereof granted under Clause 44 or by any revision thereof agreed under Clause 46(3) as the case may be.

(b) If the Contractor fails to achieve substantial completion of the whole of the Works within the time so prescribed he shall pay to the Employer the said sum for every week or day (as the case may be) which shall elapse between the date on which the prescribed time expired and the date the whole of the Works is substantially completed.

Provided that if any part of the Works is certified as substantially complete pursuant to Clause 48 before the completion of the whole of the Works the said sum shall be reduced by the proportion which the value of the part so completed bears to the value of the whole of the Works.

Liquidated damages for delay in substantial completion where the whole of the Works is divided into Sections

(2) (a) Where the Works is divided into Sections (together comprising the whole of the Works) which are required to be completed within particular times as stated in the Appendix to the Form of Tender sub-clause (1) of this Clause shall not apply and the said Appendix shall include a sum in respect of each Section which represents the Employer's genuine pre-estimate (expressed per week or per day as the case may be) of the damages likely to be suffered by him if that Section is not substantially completed within the time prescribed by Clause 43 or by any extension thereof granted under Clause 44 or by any revision thereof agreed under Clause 46(3) as the case may be.

(b) If the Contractor fails to achieve substantial completion of any Section within the time so prescribed he shall pay to the Employer the appropriate stated sum for every week or day (as the case may be) which shall elapse between the date on which the prescribed time expired and the date of substantial completion of that Section.

Provided that if any part of that Section is certified as substantially complete pursuant to Clause 48 before the completion of the whole thereof the appropriate stated sum shall be reduced by the proportion which the value of the part so completed bears to the value of the whole of that Section.

(c) Liquidated damages in respect of two or more Sections may where circumstances so dictate run concurrently.

Damages not a penalty

(3) All sums payable by the Contractor to the Employer pursuant to this Clause shall be paid as liquidated damages for delay and not as a penalty.

Limitation of liquidated damages

(4) (a) The total amount of liquidated damages in respect of the whole of the Works or any Section thereof shall be limited to the appropriate sum stated in the Appendix to the Form of Tender. If no such limit is stated therein then liquidated damages without limit shall apply.

(b) Should there be omitted from the Appendix to the Form of Tender any sum required to be inserted therein either by sub-clause (1)(a) or by sub-clause (2)(a) of this Clause as the case may be or if any such sum is stated to be "nil" then to that extent damages shall not be payable.

Recovery and reimbursement of liquidated damages

(5) The Employer may

(a) deduct and retain the amount of any liquidated damages becoming due under the provision of this Clause from any sums due or which become due to the Contractor or

(b) require the Contractor to pay such amount to the Employer forthwith.

If upon a subsequent or final review of the circumstances causing delay the Engineer grants a relevant extension or further extension of time the Employer shall no longer be entitled to liquidated damages in respect of the period of such extension.

Any sum in respect of such period which may already have been recovered under this Clause shall be reimbursed forthwith to the Contractor together with interest compounded monthly at the rate provided for in Clause 60(7) from the date on which such sums were recovered from the Contractor.

Intervention of variations etc.

(6) If after liquidated damages have become payable in respect of any part of the Works the Engineer orders a variation under Clause 51 or adverse physical conditions or artificial obstructions within the meaning of Clause 12 are encountered or any other situation outside the Contractor's control arises any of which in the Engineer's opinion results in further delay to that part of the Works

(a) the Engineer shall so notify the Contractor and the Employer in writing and

(b) the Employer's further entitlement to liquidated damages in respect of that part of the Works shall be suspended until the Engineer notifies the Contractor and the Employer in writing that the further delay has come to an end.

Such suspension shall not invalidate any entitlement to liquidated damages which accrued before the period of further delay started to run and subject to any subsequent or final review of the circumstances causing delay any monies already deducted or paid as liquidated damages under the provision of this Clause may be retained by the Employer.

CERTIFICATE OF SUBSTANTIAL COMPLETION

Notification of substantial completion

48 (1) When the Contractor considers that

(a) the whole of the Works or

(b) any Section in respect of which a separate time for completion is provided in the Appendix to the Form of Tender

has been substantially completed and has satisfactorily passed any final test that may be prescribed by the Contract he may give notice in writing to that effect to the Engineer or to the Engineer's Representative. Such notice shall be accompanied by an undertaking to finish any outstanding work in accordance with the provisions of Clause 49(1).

Certification of substantial completion

(2) The Engineer shall within 21 days of the date of delivery of such notice either

(a) issue to the Contractor (with a copy to the Employer) a Certificate of Substantial Completion stating the date on which in his opinion the Works were or the Section was substantially completed in accordance with the Contract or

(b) give instructions in writing to the Contractor specifying all the work which in the Engineer's opinion requires to be done by the Contractor before the issue of such certificate.

If the Engineer gives such instructions the Contractor shall be entitled to receive a Certificate of Substantial Completion within 21 days of completion to the satisfaction of the Engineer of the work specified in the said instructions.

Premature use by Employer

(3) If any substantial part of the Works has been occupied or used by the Employer other than as provided in the Contract the Contractor may request in writing and the Engineer shall issue a Certificate of Substantial Completion in respect thereof. Such certificate shall take effect from the date of delivery of the Contractor's request and upon the issue of such certificate the Contractor shall be deemed to have undertaken to complete any outstanding work in that part of the Works during the Defects Correction Period.

Substantial completion of other parts of the Works

(4) If the Engineer considers that any part of the Works has been substantially completed and has passed any final test that may be prescribed by the Contract he may issue a Certificate of Substantial Completion in respect of that part of the Works before completion of the whole of the Works and upon the issue of such certificate the Contractor shall be deemed to have undertaken to complete any outstanding work in that part of the Works during the Defects Correction Period.

Reinstatement of ground

(5) A Certificate of Substantial Completion given in respect of any Section or part of the Works before completion of the whole shall not be deemed to certify completion of any ground or surfaces requiring reinstatement unless such certificate shall expressly so state.

OUTSTANDING WORK AND DEFECTS

Work outstanding **49** (1) The undertaking to be given under Clause 48(1) may after agreement between the Engineer and the Contractor specify a time or times within which the outstanding work shall be completed. If no such times are specified any outstanding work shall be completed as soon as practicable during the Defects Correction Period.

Carrying out of work of repair etc. (2) The Contractor shall deliver up to the Employer the Works and each Section and part thereof at or as soon as practicable after the end of the relevant Defects Correction Period in the condition required by the Contract (fair wear and tear excepted) to the satisfaction of the Engineer. To this end the Contractor shall as soon as practicable carry out all work of repair amendment reconstruction rectification and making good of defects of whatever nature as may be required of him in writing by the Engineer during the relevant Defects Correction Period or within 14 days after its expiry as a result of an inspection made by or on behalf of the Engineer prior to its expiry.

Cost of work of repair etc. (3) All work required under sub-clause (2) of this Clause shall be carried out by the Contractor at his own expense if in the Engineer's opinion it is necessary due to the use of materials or workmanship not in accordance with the Contract or to neglect or failure by the Contractor to comply with any of his obligations under the Contract. In any other event the value of such work shall be ascertained and paid for as if it were additional work.

Remedy on Contractor's failure to carry out work required (4) If the Contractor fails to do any such work as aforesaid the Employer shall be entitled to carry out that work by his own workpeople or by other contractors and if it is work which the Contractor should have carried out at his own expense the Employer shall be entitled to recover the cost thereof from the Contractor and may deduct the same from any monies that are or may become due to the Contractor.

Contractor to search **50** The Contractor shall if required by the Engineer in writing carry out such searches tests or trials as may be necessary to determine the cause of any defect imperfection or fault under the directions of the Engineer. Unless the defect imperfection or fault is one for which the Contractor is liable under the Contract the cost of the work carried out by the Contractor as aforesaid shall be borne by the Employer. If the defect imperfection or fault is one for which the Contractor is liable the cost of the work carried out as aforesaid shall be borne by the Contractor and he shall in such case repair rectify and make good such defect imperfection or fault at his own expense in accordance with Clause 49.

ALTERATIONS, ADDITIONS AND OMISSIONS

Ordered variations **51** (1) The Engineer

(a) shall order any variation to any part of the Works that is in his opinion necessary for the completion of the Works and

(b) may order any variation that for any other reason shall in his opinion be desirable for the completion and/or improved functioning of the Works.

Such variations may include additions omissions substitutions alterations changes in quality form character kind position dimension level or line and changes in any specified sequence method or timing of construction required by the Contract and may be ordered during the Defects Correction Period.

Ordered variations to be in writing

(2) All variations shall be ordered in writing but the provisions of Clause 2(6) in respect of oral instructions shall apply.

Variations proposed by the Contractor

(3) The Contractor may at any time submit to the Engineer proposals to vary the drawings and specification. If such proposals are accepted (whether as submitted or with modifications) the Engineer shall issue an appropriate order under sub-clause (1) of this Clause.

Variation not to affect Contract

(4) No variation ordered in accordance with sub-clauses (1) and (2) of this Clause shall in any way vitiate or invalidate the Contract but the value (if any) of all such variations shall be taken into account in ascertaining the amount of the Contract Price except to the extent that such variation is necessitated by the Contractor's default.

Changes in quantities

(5) No order in writing shall be required for increase or decrease in the quantity of any work where such increase or decrease is not the result of an order given under this Clause but is the result of the quantities exceeding or being less than those stated in the Bill of Quantities.

Valuation of ordered variations

52 (1) If requested by the Engineer the Contractor shall submit his quotation for any proposed variation and his estimate of any consequential delay. Wherever possible the value and delay consequences (if any) of each variation shall be agreed before the order is issued or before work starts.

Sharing the consequences

(2) Where the varied work results from a Contractor's proposal pursuant to Clause 51(3) the consequences (if any) as to delay acceleration or valuation shall be shared between the Employer and the Contractor having regard to the particular circumstances of each case.

(3) Where a request is not made or agreement is not reached under sub-clause (1) of this Clause the valuation of variations ordered by the Engineer in accordance with Clause 51 shall be ascertained as follows.

(a) As soon as possible after receipt of the variation the Contractor shall submit to the Engineer

(i) his quotation for any extra or substituted works necessitated by the variation having due regard to any rates or prices included in the Contract and

(ii) his estimate of any delay occasioned thereby and

(iii) his estimate of the cost of any such delay.

(b) Within 14 days of receiving the said submissions the Engineer shall

(i) accept those submissions or

(ii) negotiate with the Contractor thereon.

(c) Upon reaching agreement with the Contractor the Contract Price shall be amended accordingly.

(4) Failing agreement between the Engineer and the Contractor under either sub-clause (1) or (2) the value of variations ordered by the Engineer in accordance with Clause 51 shall be ascertained by the Engineer in accordance with the following principles and be notified to the Contractor.

(a) Where work is of similar character and carried out under similar conditions to work priced in the Bill of Quantities it shall be valued at such rates and prices contained therein as may be applicable.

(b) Where work is not of a similar character or is not carried out under similar conditions or is ordered during the Defects Correction Period the rates and prices in the Bill of Quantities shall be used as the basis for valuation so far as may be reasonable failing which a fair valuation shall be made.

Engineer to fix rates

(5) If in the opinion of the Engineer or the Contractor any rate or price contained in the Contract for any item of work (not being the subject of any variation) is by reason of any variation rendered unreasonable or inapplicable either the Engineer shall give to the Contractor or the Contractor shall give to the Engineer notice before the varied work is commenced or as soon thereafter as is reasonable in all the circumstances that such rate or price should be increased or decreased and the Engineer shall fix such rate or price as in the circumstances he shall think reasonable and proper and shall so notify the Contractor.

Daywork

(6) The Engineer may if in his opinion it is necessary or desirable order in writing that any additional or substituted work shall be carried out on a daywork basis in accordance with the provisions of Clause 56(4).

Additional payments

53 (1) If the Contractor intends to claim a higher rate or price than one notified to him by the Engineer pursuant to sub-clauses (3) and (4) of Clause 52 or Clause 56(2) the Contractor shall within 28 days after such notification give notice in writing of his intention to the Engineer.

(2) If the Contractor intends to claim any additional payment pursuant to any Clause of these Conditions other than sub-clauses (3) and (4) of Clause 52 or Clause 56(2) he shall give notice in writing of his intention to the Engineer as soon as may be reasonable and in any event within 28 days after the happening of the events giving rise to the claim.

Upon the happening of such events the Contractor shall keep such contemporary records as may reasonably be necessary to support any claim he may subsequently wish to make.

(3) Without necessarily admitting the Employer's liability the Engineer may upon receipt of a notice under this Clause instruct the Contractor to keep such contemporary records or further contemporary records as the case may be as are reasonable and may be material to the claim of which notice has been given and the Contractor shall keep such records.

The Contractor shall permit the Engineer to inspect all records kept pursuant to Clause 53 and shall supply him with copies thereof as and when the Engineer shall so instruct.

(4) After the giving of a notice to the Engineer under this Clause the Contractor shall as soon as is reasonable in all the circumstances send to the Engineer a first interim account giving full and detailed particulars of the amount claimed to that date and of the grounds upon which the claim is based.

Thereafter at such intervals as the Engineer may reasonably require the Contractor shall send to the Engineer further up to date accounts giving the accumulated total of the claim and any further grounds upon which it is based.

(5) If the Contractor fails to comply with any of the provisions of this Clause in respect of any claim which he shall seek to make then the Contractor shall be entitled to payment in respect thereof only to the extent that the Engineer has not been prevented from or substantially prejudiced by such failure in investigating the said claim.

(6) The Contractor shall be entitled to have included in any interim payment certified by the Engineer pursuant to Clause 60 such amount in respect of any claim as the Engineer may consider due to the Contractor provided that the Contractor shall have supplied sufficient particulars to enable the Engineer to determine the amount due.

If such particulars are insufficient to substantiate the whole of the claim the Contractor shall be entitled to payment in respect of such part of the claim as the particulars may substantiate to the satisfaction of the Engineer.

PROPERTY IN MATERIALS AND CONTRACTOR'S EQUIPMENT

Non-removal of Materials and Contractor's Equipment

54 (1) No Contractor's Equipment Temporary Works materials for Temporary Works or other goods or materials owned by the Contractor and brought on to the Site for the purposes of the Contract shall be removed without the written consent of the Engineer which consent shall not unreasonably be withheld.

Liability for loss or damage to Contractor's Equipment

(2) The Employer shall not at any time be liable save as mentioned in Clause 20(2) for the loss of or damage to any Contractor's Equipment Temporary Works goods or materials.

Disposal of Contractor's Equipment

(3) If the Contractor fails to remove any of the said Contractor's Equipment Temporary Works goods or materials as required by Clause 33 within such reasonable time after completion of the Works as the Engineer may allow then the Employer may sell or otherwise dispose of such items. From the proceeds of the sale of any such items the Employer shall be entitled to retain any costs or expenses incurred in connection with their sale and disposal before paying the balance (if any) to the Contractor.

Vesting of goods and materials not on Site

(4) With a view to securing payment under Clause 60(1)(c) the Contractor may (and shall if the Engineer so directs) transfer to the Employer the property in goods and materials listed in the Appendix to the Form of Tender or as subsequently agreed between the Contractor and the Employer before the same are delivered to the Site provided that the goods and materials

(a) have been manufactured or prepared and are substantially ready for incorporation in the Works and

(b) are the property of the Contractor or the contract for the supply of the same expressly provides that the property therein shall pass unconditionally to the Contractor upon the Contractor taking the action referred to in sub-clause (5) of this Clause and

(c) have been marked and set aside in accordance with sub-clause (5) of this Clause.

Action by Contractor

(5) The intention of the Contractor to transfer the property in any goods or materials to the Employer in accordance with this Clause shall be evidenced by the Contractor taking or causing the supplier of those goods or materials to take the following actions.

(a) Provide to the Engineer documentary evidence that the property of the said goods or materials has vested in the Contractor.

(b) Suitably mark or otherwise plainly identify the goods and materials so as to show that their destination is the Site that they are the property of the Employer and (where they are not stored at the premises of the Contractor) to whose order they are held.

(c) Set aside and store the said goods and materials so marked and identified to the satisfaction of the Engineer.

(d) Send to the Engineer a schedule listing and giving the value of every item of the goods and materials so set aside and stored and inviting him to inspect them.

Vesting in the Employer

(6) Upon the Engineer approving in writing the transfer in ownership of any goods and materials for the purposes of this Clause they shall vest in and become the absolute property of the Employer and thereafter shall be in the possession of the Contractor for the sole purpose of delivering them to the Employer and incorporating them in the Works and shall not be within the ownership control or disposition of the Contractor. Provided always that

(a) approval by the Engineer for the purposes of this Clause or any payment certified by him in respect of goods and materials pursuant to Clause 60 shall be without prejudice to the exercise of any power of the Engineer contained in this Contract to reject any goods or materials which are not in accordance with the provisions of the Contract and upon any such rejection the property in the rejected goods or materials shall immediately re-vest in the Contractor and

(b) the Contractor shall be responsible for any loss or damage to such goods or materials and for the cost of storing handling and transporting the same and shall effect such additional insurance as may be necessary to cover the risk of such loss or damage from any cause.

Lien on goods and materials

(7) Neither the Contractor nor a sub-contractor nor any person shall have a lien on any goods or materials which have vested in the Employer under sub-clause (6) of this Clause for any sum due to the Contractor sub-contractor or other person and the Contractor shall take all steps reasonably necessary to ensure that the title of the Employer and the exclusion of any such lien are brought to the notice of sub-contractors and other persons dealing with such goods or materials.

Delivery to the Employer of vested goods or materials

(8) Upon cessation of the employment of the Contractor under this Contract before the completion of the Works whether as a result of the operation of Clause 63 64 or 65 or otherwise the Contractor shall deliver to the Employer any goods or materials the property in which has vested in the Employer by virtue of sub-clause (6) of this Clause and if he fails to do so the Employer may enter any premises of the Contractor or of any sub-contractor and remove such goods and materials and recover the cost of doing so from the Contractor.

Incorporation in sub-contracts

(9) The Contractor shall incorporate provisions equivalent to those provided in this Clause in every sub-contract in which provision is to be made for the payment in respect of goods or materials before the same have been delivered to the Site.

MEASUREMENT

Quantities

55 (1) The quantities set out in the Bill of Quantities are the estimated quantities of the work but they are not to be taken as the actual and correct quantities of the Works to be carried out by the Contractor in fulfilment of his obligations under the Contract.

Correction of errors

(2) No error in description in the Bill of Quantities or omission therefrom shall vitiate the Contract nor release the Contractor from the carrying out of the whole or any part of the Works according to the Drawings and Specification or from any of his obligations or liabilities under the Contract. Any such error or omission shall be corrected by the Engineer and the value of the work actually carried out shall be ascertained in accordance with Clause 52(2) or (3). Provided that there shall be no rectification of any errors omissions or wrong estimates in the descriptions rates and prices inserted by the Contractor in the Bill of Quantities.

Measurement and valuation

56 (1) The Engineer shall except as otherwise stated ascertain and determine by admeasurement the value in accordance with the Contract of the work done in accordance with the Contract.

Increase or decrease of rate

(2) Should the actual quantities carried out in respect of any item be greater or less than those stated in the Bill of Quantities and if in the opinion of the Engineer such increase or decrease of itself shall so warrant the Engineer shall after consultation with the Contractor determine an appropriate increase or decrease of any rates or prices rendered unreasonable or inapplicable in consequence thereof and shall notify the Contractor accordingly.

Attending for measurement

(3) The Engineer shall when he requires any part or parts of the work to be measured give reasonable notice to the Contractor who shall attend or send a qualified agent to assist the Engineer or the Engineer's Representative in making such measurement and shall furnish all particulars required by either of them. Should the Contractor not attend or neglect or omit to send such agent then the measurement made by the Engineer or approved by him shall be taken to be the correct measurement of the work.

Daywork

(4) Where any work is carried out on a daywork basis the Contractor shall be paid for such work under the conditions and at the rates and prices set out in the daywork schedule included in the Contract or failing the inclusion of a daywork schedule he shall be paid at the rates and prices and under the conditions contained in the "Schedules of Dayworks carried out incidental to Contract Work" issued by The Civil Engineering Contractors Association (formerly issued by The Federation of Civil Engineering Contractors) current at the date of the carrying out of the daywork.

The Contractor shall furnish to the Engineer such records receipts and other documentation as may be necessary to prove amounts paid and/or costs incurred. Such returns shall be in the form and delivered at the times the Engineer shall direct and shall be agreed within a reasonable time.

Before ordering materials the Contractor shall if so required submit to the Engineer quotations for the same for his approval.

Method of measurement

57 Unless otherwise provided in the Contract or unless general or detailed description of the work in the Bill of Quantities or any other statement clearly shows to the contrary the Bill of Quantities shall be deemed to have been prepared and measurements shall be made according to the procedure set out in the "Civil Engineering Standard Method of Measurement Third Edition 1991" approved by the Institution of Civil Engineers and the Federation of Civil Engineering Contractors in association with the Association of Consulting Engineers or such later or amended edition thereof as may be stated in the Appendix to the Form of Tender to have been adopted in its preparation.

PROVISIONAL AND PRIME COST SUMS AND NOMINATED SUB-CONTRACTS

Use of Provisional Sums

58 (1) In respect of every Provisional Sum the Engineer may order either or both of the following.

(a) Work to be carried out or goods materials or services to be supplied by the Contractor the value thereof being determined in accordance with Clause 52 and included in the Contract Price.

(b) Work to be carried out or goods materials or services to be supplied by a Nominated Sub-contractor in accordance with Clause 59.

Use of Prime Cost Items

(2) In respect of every Prime Cost Item the Engineer may order either or both of the following.

(a) Subject to Clause 59 that the Contractor employ a sub-contractor nominated by the Engineer for the carrying out of any work or the supply of any goods materials or services included therein.

(b) With the consent of the Contractor that the Contractor himself carry out any such work or supply any such goods materials or services in which event the Contractor shall be paid in accordance with the terms of a quotation submitted by him and accepted by the Engineer or in the absence thereof the value shall be determined in accordance with Clause 52 and included in the Contract Price.

Design requirements to be expressly stated

(3) If in connection with any Provisional Sum or Prime Cost Item the services to be provided include any matter of design or specification of any part of the Permanent Works or of any equipment or plant to be incorporated therein such requirement shall be expressly stated in the Contract and shall be included in any Nominated Sub-contract. The obligation of the Contractor in respect thereof shall be only that which has been expressly stated in accordance with this sub-clause.

Nominated Sub-contractors — objection to nomination

59 (1) The Contractor shall not be under any obligation to enter into a sub-contract with any Nominated Sub-contractor against whom the Contractor may raise reasonable objection or who declines to enter into a sub-contract with the Contractor containing provisions

(a) that in respect of the work goods materials or services the subject of the sub-contract the Nominated Sub-contractor will undertake towards the Contractor such obligations and liabilities as will enable the Contractor to discharge his own obligations and liabilities towards the Employer under the terms of the Contract

(b) that the Nominated Sub-contractor will indemnify and keep indemnified the Contractor against all claims demands and proceedings damages costs charges and expenses whatsoever arising out of or in connection with any failure by the Nominated Sub-contractor to perform such obligations or fulfil such liabilities

(c) that the Nominated Sub-contractor will indemnify and keep indemnified the Contractor from and against any negligence by the Nominated Sub-contractor his agents workmen and servants and against any misuse by him or them of any Contractor's Equipment or Temporary Works provided by the Contractor for the purposes of the Contract and for all claims as aforesaid

(d) that the Nominated Sub-contractor will provide the Contractor with security for the proper performance of the sub-contract and

(e) equivalent to those contained in Clause 65.

Engineer's action upon objection to nomination or upon determination of Nominated Sub-contract

(2) If pursuant to sub-clause (1) of this Clause the Contractor declines to enter into a sub-contract with a sub-contractor nominated by the Engineer or if during the course of the Nominated Sub-contract the Contractor shall validly terminate the employment of the Nominated Sub-contractor as a result of his default the Engineer shall

(a) nominate an alternative sub-contractor in which case sub-clause (1) of this Clause shall apply or

(b) by order under Clause 51 vary the Works or the work goods materials or services in question or

(c) by order under Clause 51 omit any or any part of such works goods materials or services so that they may be provided by workmen contractors or suppliers employed by the Employer either

(i) concurrently with the Works (in which case Clause 31 shall apply) or

(ii) at some other date

and in either case there shall nevertheless be included in the Contract Price such sum (if any) in respect of the Contractor's charges and profit being a percentage of the estimated value of such omission as would have been payable had there been no such omission and the value thereof had been that entered in the Bill of Quantities or inserted in the Appendix to the Form of Tender as the case may be or

(d) instruct the Contractor to secure a sub-contractor of his own choice and to submit a quotation for the work goods materials or services in question to be so performed or provided for the Engineer's consideration and action or

(e) invite the Contractor himself to carry out or supply the work goods materials or services in question under Clause 58(1)(a) or Clause 58(2)(b) or on a daywork basis as the case may be.

Contractor responsible for Nominated Sub-contractors

(3) Except as otherwise provided in Clause 58 (3) the Contractor shall be as responsible for the work carried out or goods materials or services supplied by a Nominated Sub-contractor employed by him as if he had himself carried out such work or supplied such goods materials or services.

Nominated Sub-contractor's default

(4) (a) If any event arises which in the opinion of the Contractor justifies the exercise of his right under any forfeiture clause to terminate the sub-contract or to treat the sub-contract as repudiated by the Nominated Sub-contractor he shall at once notify the Engineer in writing giving his reasons.

Termination of Sub-contract

(b) With the consent in writing of the Engineer the Contractor may give notice to the Nominated Sub-contractor expelling him from the Sub-contract works pursuant to any forfeiture clause or rescinding the Sub-contract as the case may be. If however the Engineer's consent is withheld the Contractor shall be entitled to appropriate instructions under Clause 13.

Engineer's action upon termination

(c) In the event that the Nominated Sub-contractor is expelled from the Sub-contract works the Engineer shall at once take such action as is required under sub-clause (2) of this Clause.

Recovery of additional expense

(d) Having with the Engineer's consent terminated the Nominated Sub-contract the Contractor shall take all necessary steps and proceedings as are available to him to recover all additional expenses that are incurred from the Sub-contractor or under the security provided pursuant to sub-clause (1)(d) of this Clause. Such expenses shall include any additional expenses incurred by the Employer as a result of the termination.

Reimbursement of Contractor's loss

(e) If and to the extent that the Contractor fails to recover all his reasonable expenses of completing the Sub-contract works and all his proper additional expenses arising from the termination the Employer will reimburse the Contractor his unrecovered expenses.

Consequent delay

(f) The Engineer shall take any delay to the completion of the Works consequent upon the Nominated Sub-contractor's default into account in determining any extension of time to which the Contractor is entitled under Clause 44.

Provisions for payment

(5) For all work carried out or goods materials or services supplied by Nominated Sub-contractors there shall be included in the Contract Price

(a) the actual price paid or due to be paid by the Contractor in accordance with the terms of the sub-contract (unless and to the extent that any such payment is the result of a default of the Contractor) net of all trade and other discounts, rebates and allowances other than any discount obtainable by the Contractor for prompt payment

(b) the sum (if any) provided in the Bill of Quantities for labours in connection therewith and

(c) in respect of all other charges and profit a sum being a percentage of the actual price paid or due to be paid calculated (where provision has been made in the Bill of Quantities for a rate to be set against the relevant item of prime cost) at the rate inserted by the Contractor against that item or (where no such provision has been made) at the rate inserted by the Contractor in the Appendix to the Form of Tender as the percentage for adjustment of sums set against Prime Cost Items.

Production of vouchers etc.

(6) The Contractor shall when required by the Engineer produce all quotations invoices vouchers sub-contract documents accounts and receipts in connection with expenditure in respect of work carried out by all Nominated Sub-contractors.

Payment to Nominated Sub-contractors

(7) Before issuing any certificate under Clause 60 the Engineer shall be entitled to demand from the Contractor reasonable proof that all sums (less retentions provided for in the Sub-contract) included in previous certificates in respect of the work carried out or goods or materials or services supplied by Nominated Sub-contractors have been paid to the Nominated Sub-contractors or discharged by the Contractor in default whereof unless the Contractor shall

(a) give details to the Engineer in writing of any reasonable cause he may have for withholding or refusing to make such payment and

(b) produce to the Engineer reasonable proof that he has so informed such Nominated Sub-contractor in writing

the Employer shall be entitled to pay to such Nominated Sub-contractor direct upon the certification of the Engineer all payments (less retentions provided for in the Sub-contract) which the Contractor has failed to make to such Nominated Sub-contractor and to deduct by way of set-off the amount so paid by the Employer from any sums due or which become due from the Employer to the Contractor. Provided always that where the Engineer has certified and the Employer has made direct payment to the Nominated Sub-contractor the Engineer shall in issuing any further certificate in favour of the Contractor deduct from the amount thereof the amount so paid but shall not withhold or delay the issue of the certificate itself when due to be issued under the terms of the Contract.

CERTIFICATES AND PAYMENT

Monthly statements **60**

(1) Unless otherwise agreed the Contractor shall submit to the Engineer at monthly intervals commencing one month after the Works Commencement Date a statement (in such form if any as may be prescribed in the Specification) showing

(a) the estimated contract value of the Permanent Works carried out up to the end of that month

(b) a list of any goods or materials delivered to the Site for but not yet incorporated in the Permanent Works and their value

(c) a list of any of those goods or materials identified in the Appendix to the Form of Tender which have not yet been delivered to the Site but of which the property has vested in the Employer pursuant to Clause 54 and their value and

(d) the estimated amounts to which the Contractor considers himself entitled in connection with all other matters for which provision is made under the Contract including any Temporary Works or Contractor's Equipment for which separate amounts are included in the Bill of Quantities

unless in the opinion of the Contractor such values and amounts together will not justify the issue of an interim certificate.

Amounts payable in respect of Nominated Sub-contracts are to be listed separately.

Monthly payments

(2) Within 25 days of the date of delivery of the Contractor's monthly statement to the Engineer or the Engineer's Representative in accordance with sub-clause (1) of this Clause the Engineer shall certify and within 28 days of the same date the Employer shall pay to the Contractor (after deducting any previous payments on account)

(a) the amount which in the opinion of the Engineer on the basis of the monthly statement is due to the Contractor on account of sub-clauses (1)(a) and (1)(d) of this Clause less a retention as provided in sub-clause (5) of this Clause and

(b) such amounts (if any) as the Engineer may consider proper (but in no case exceeding the percentage of the value stated in the Appendix to the Form of Tender) in respect of sub-clauses (1)(b) and (1)(c) of this Clause.

The payments become due on certification with the final date for payment being 28 days after the date of delivery of the Contractor's monthly statement.

The amounts certified in respect of Nominated Sub-contracts shall be shown separately in the certificate.

Minimum amount of certificate

(3) Until the whole of the Works has been certified as substantially complete in accordance with Clause 48 the Engineer shall not be bound to issue an interim certificate for a sum less than that stated in the Appendix to the Form of Tender but thereafter he shall be bound to do so and the certification and payment of amounts due to the Contractor shall be in accordance with the time limits contained in this Clause.

Final account

(4) Not later than 3 months after the date of the Defects Correction Certificate the Contractor shall submit to the Engineer a statement of final account and supporting documentation showing in detail the value in accordance with the Contract of the Works carried out together with all further sums which the Contractor considers to be due to him under the Contract up to the date of the Defects Correction Certificate.

Within 3 months after receipt of this final account and of all information reasonably required for its verification the Engineer shall issue a certificate stating the amount which in his opinion is finally due under the Contract from the Employer to the Contractor or from the Contractor to the Employer as the case may be up to the date of the Defects Correction Certificate and after giving credit to the Employer for all amounts previously paid by the Employer and for all sums to which the Employer is entitled under the Contract.

Such amount shall subject to Clause 47 be paid to or by the Contractor as the case may require. The payment becomes due on certification. The final date for payment is 28 days later.

Retention

(5) The retention to be made pursuant to sub-clause (2)(a) of this Clause shall be the difference between

(a) an amount calculated at the rate indicated in and up to the limit set out in the Appendix to the Form of Tender upon the amount due to the Contractor on account of sub-clauses (1)(a) and (1)(d) of this Clause and

(b) any payment which shall have become due under sub-clause (6) of this Clause.

Payment of retention

(6) (a) Upon the issue of a Certificate of Substantial Completion in respect of any Section or part of the Works there shall become due to the Contractor one half of such proportion of the amount calculated to date under sub-clause (5)(a) of this Clause as the value of the Section or part bears to the value of the whole of the Works completed to date as certified under sub-clause (2)(a) of this Clause.

The total of the payments which shall become due under this sub-clause shall in no event exceed one half of the limit of retention set out in the Appendix to the Form of Tender.

(b) Upon issue of the Certificate of Substantial Completion in respect of the whole of the Works there shall become due to the Contractor one half of the amount calculated in accordance with sub-clause (5)(a) of this Clause less any payments which shall have become due under sub-clause (6)(a) of this Clause. Within 10 days of the date of issue of the said Certificate the Engineer shall certify the amount due.

Payment becomes due on certification of the amount due with the final date for payment being 14 days after the issue of the Certificate of Substantial Completion.

(c) At the end of the Defects Correction Period or if more than one the last of such periods there shall become due to the Contractor the remainder of the retention money. Within 10 days of the date of the end of the said period the Engineer shall certify the amount due.

Payment becomes due on certification of the amount due with the final date for payment being 14 days after the end of the said period notwithstanding that at that time there may be outstanding claims by the Contractor against the Employer.

Provided that if at that time there remains to be carried out by the Contractor any outstanding work referred to under Clause 48 or any work ordered pursuant to Clauses 49 or 50 the Engineer may withhold certification until the completion of such work of so much of the said remainder as shall in the opinion of the Engineer represent the cost of the work remaining to be carried out.

Interest on overdue payments

(7) In the event of

(a) failure by the Engineer to certify or the Employer to make payment in accordance with sub-clauses (2) (4) or (6) of this Clause or

(b) any decision of an adjudicator or any finding of an arbitrator to such effect

the Employer shall pay to the Contractor interest compounded monthly for each day on which any payment is overdue or which should have been certified and paid at a rate equivalent to 2% per annum above the base lending rate of the bank specified in the Appendix to the Form of Tender.

If in an arbitration pursuant to Clause 66 the arbitrator holds that any sum or additional sum should have been certified by a particular date in accordance with the aforementioned sub-clauses but was not so certified this shall be regarded for the purposes of this sub-clause as a failure to certify such sum or additional sum. Such sum or additional sum shall be regarded as overdue for payment 28 days after the date by which the arbitrator holds that the Engineer should have certified the sum or if no such date is identified by the arbitrator shall be regarded as overdue for payment from the date of the Certificate of Substantial Completion for the whole of the Works.

Correction and withholding of certificates

(8) The Engineer shall have power to omit from any certificate the value of any work done goods or materials supplied or services rendered with which he may for the time being be dissatisfied and for that purpose or for any other reason which to him may seem proper may by any certificate delete correct or modify any sum previously certified by him. Provided that

(a) the Engineer shall not in any interim certificate delete or reduce any sum previously certified in respect of work done goods or materials supplied or services rendered by a Nominated Sub-contractor if the Contractor shall have already paid or be bound to pay that sum to the Nominated Sub-contractor and

(b) if the Engineer in the final certificate shall delete or reduce any sum previously certified in respect of work done goods or materials supplied or services rendered by a Nominated Sub-contractor which sum shall have been already paid by the Contractor to the Nominated Sub-contractor the Employer shall reimburse to the Contractor the amount of any sum overpaid by the Contractor to the Sub-contractor in accordance with the certificates issued under sub-clause (2) of this Clause which the Contractor shall be unable to recover from the Nominated Sub-contractor together with interest thereon at the rate stated in sub-clause (7) of this Clause from 28 days after the date of the final certificate issued under sub-clause (4) of this Clause until the date of such reimbursement.

Certificates and payment notices

(9) Every certificate issued by the Engineer pursuant to this Clause shall be sent to the Employer and on the Employer's behalf to the Contractor. By this certificate the Employer shall give notice to the Contractor specifying the amount (if any) of the payment proposed to be made and the basis on which it was calculated.

Notice of intention to withhold payment

(10) Where a payment under Clause 60(2) (4) or (6) is to differ from that certified or the Employer is to withhold payment after the final date for payment of a sum due under the Contract the Employer shall notify the Contractor in writing not less than one day before the final date for payment specifying the amount proposed to be withheld and the ground for withholding payment or if there is more than one ground each ground and the amount attributable to it.

Defects Correction Certificate

61 (1) At the end of the Defects Correction Period or if more than one the last of such periods and when all outstanding work referred to under Clause 48 and all work of repair amendment reconstruction rectification and making good of defects imperfections shrinkages and other faults referred to under Clauses 49 and 50 have been completed the Engineer shall issue to the Employer (with a copy to the Contractor) a Defects Correction Certificate stating the date on which the Contractor shall have completed his obligations to construct and complete the Works to the Engineer's satisfaction.

Unfulfilled obligations

(2) The issue of the Defects Correction Certificate shall not be taken as relieving either the Contractor or the Employer from any liability the one towards the other arising out of or in any way connected with the performance of their respective obligations under the Contract.

REMEDIES AND POWERS

Urgent repairs

62 If in the opinion of the Engineer any remedial or other work or repair is urgently necessary by reason of any accident or failure or other event occurring to in or in connection with the Works or any part thereof either during the carrying out of the Works or during the Defects Correction Period the Engineer shall so inform the Contractor with confirmation in writing.

Thereafter if the Contractor is unable or unwilling to carry out such work or repair at once the Employer may himself carry out the said work or repair using his own or other workpeople.

If the work or repair so carried out by the Employer is work which in the opinion of the Engineer the Contractor was liable to carry out at his own expense under the Contract all costs and charges properly incurred by the Employer in so doing shall on demand be paid by the Contractor to the Employer or may be deducted by the Employer from any monies due or which may become due to the Contractor.

Frustration **63** (1) If any circumstance outside the control of both parties arises during the currency of the Contract which renders it impossible or illegal for either party to fulfil his contractual obligations the Works shall be deemed to be abandoned upon the service by one party upon the other of written notice to that effect.

War clause

(2) If during the currency of the Contract there is an outbreak of war (whether war is declared or not) in which Great Britain is engaged on a scale involving general mobilization of the armed forces of the Crown

(a) the Contractor shall for a period of 28 days reckoned from midnight on the date that the order for general mobilization is given continue so far as is physically possible to carry out the Works in accordance with the Contract and

(b) if substantial completion of the whole of the Works is not achieved before the said period of 28 days has expired the Works shall thereupon be deemed to be abandoned unless the parties otherwise agree.

Removal of Contractor's Equipment

(3) Upon abandonment of the Works pursuant to sub-clauses (1) or (2)(b) of this Clause the Contractor shall with all reasonable dispatch remove from the Site all Contractor's Equipment.

In the event of any failure so to do the Employer shall have like powers to those contained in clause 54(3) to dispose of any Contractor's Equipment.

Payment on abandonment

(4) Upon abandonment of the Works pursuant to sub-clauses (1) or (2)(b) of this Clause the Employer shall pay the Contractor (in so far as such amounts or items have not already been covered by payments on account made to the Contractor) the Contract value of all work carried out prior to the date of abandonment and in addition

(a) the amounts payable in respect of any preliminary items so far as the work or service comprised therein has been carried out or performed and a proper proportion of any such items which have been partially carried out or performed

(b) the cost of materials or goods reasonably ordered for the Works which have been delivered to the Contractor or of which the Contractor is legally liable to accept delivery (such materials or goods becoming the property of the Employer upon such payment being made to the Contractor)

(c) a sum being the amount of any expenditure reasonably incurred by the Contractor in the expectation of completing the whole of the Works insofar as such expenditure has not been recovered by any other payments referred to in this sub-clause and

(d) the reasonable cost of removal under sub-clause (3) of this Clause.

To this end and without prejudice to the provisions of sub-clause (5) of this Clause the provisions of Clause 60(4) shall apply to this sub-clause as if the date of abandonment was the date of issue of the Defects Correction Certificate.

Works substantially completed

(5) If upon abandonment of the Works any Section or part of the Works has been substantially completed in accordance with Clause 48 or is completed so far as to be usable then in connection therewith

(a) the Contractor may at his discretion and in lieu of his obligations under Clauses 49 and 50 allow against the sum due to him pursuant to sub-clause (4) of this Clause the cost (calculated as at the date of abandonment) of repair rectification and making good for which he would have been liable under the said Clauses had they continued to be applicable and

(b) the Employer shall not be entitled to withhold payment under Clause 60(6)(c) of the second half of the retention money or any part thereof except such sum as the Contractor may allow under the provisions of the last preceding paragraph.

Contract to continue in force

(6) Save as aforesaid the Contract shall continue to have full force and effect.

Default of the Employer

64 (1) In the event that the Employer

(a) assigns or attempts to assign the Contract or any part thereof or any benefit or interest thereunder without the prior written consent of the Contractor or

(b) (i) becomes bankrupt or presents his petition in bankruptcy or

(ii) has a receiving order or administration order made against him or

(iii) makes an arrangement with or an assignment in favour of his creditors or

(iv) agrees to perform the Contract under a committee of inspection of his creditors or

(v) (being a corporation) has a receiver or administrator appointed or goes into liquidation (other than a voluntary liquidation for the purposes of amalgamation or reconstruction) or

(c) has an execution levied on his goods which is not stayed or discharged within 28 days

then the Contractor may after giving 7 days notice in writing to the Employer specifying the event relied on terminate his employment under the Contract without thereby avoiding the Contract or releasing the Employer from any of his obligations or liabilities under the Contract.

Provided that the Contractor may extend the period of notice to give the Employer an opportunity to remedy the situation.

Removal of Contractor's Equipment

(2) Upon expiry of the 7 days notice referred to in sub-clause (1) of this Clause and notwithstanding the provisions of Clause 54 the Contractor shall with all reasonable despatch remove from the site all Contractor's Equipment.

Payment upon termination

(3) Upon termination of the Contractor's employment pursuant to sub-clause (1) of this Clause the Employer shall be under the same obligations with regard to payment as if the Works had been abandoned under the provisions of Clause 63.

Provided that in addition to payments specified under Clause 63(4) the Employer shall pay to the Contractor the amount of any loss or damage to the Contractor arising from or as a consequence of such termination.

Default of Contractor **65**

(1) In the event that the Contractor

(a) assigns or attempts to assign the Contract or any part thereof or any benefit or interest thereunder without the prior written consent of the Employer or

(b) is in breach of Clause 4(1) or

(c) (i) becomes bankrupt or presents his petition in bankruptcy or

(ii) has a receiving order or administration order made against him or

(iii) makes an arrangement with or an assignment in favour of his creditors or

(iv) agrees to carry out the Contract under a committee of inspection of his creditors or

(v) (being a corporation) has a receiver or administrator appointed or goes into liquidation (other than a voluntary liquidation for the purposes of amalgamation or reconstruction) or

(d) has an execution levied on his goods which is not stayed or discharged within 28 days

or if the Engineer certifies in writing to the Employer with a copy to the Contractor that in his opinion the Contractor

(e) has abandoned the Contract without due cause or

(f) without reasonable excuse has failed to commence the Works in accordance with Clause 41 or

(g) has suspended the progress of the Works without due cause for 14 days after receiving from the Engineer written notice to proceed or

(h) has failed to remove goods or materials from the Site or to pull down and replace work for 14 days after receiving from the Engineer written notice that the said goods materials or work has been condemned and rejected by the Engineer or

(j) despite previous warnings by the Engineer in writing is failing to proceed with the Works with due diligence or is otherwise persistently or fundamentally in breach of his obligations under the Contract then the Employer may after giving 7 days notice in writing to the Contractor specifying the event relied on enter upon the Works and any other parts of the Site provided by the Employer and expel the Contractor therefrom without thereby avoiding the Contract or releasing the Contractor from any of his obligations or liabilities under the Contract.

Where a notice of termination is given pursuant to a certificate issued by the Engineer under this sub-clause it shall be given as soon as is reasonably possible after receipt of the certificate.

Provided that the Employer may extend the period of notice to give the Contractor an opportunity to remedy the situation.

Completing the Works

(2) Where the Employer has entered upon the Works and any other parts of the Site as set out in sub-clause (1) of this Clause he may

(a) complete the Works himself or

(b) employ any other contractor to complete the Works

and in either case may use for such completion any of the Contractor's Equipment Temporary Works goods and materials on any part of the Site. For this purpose the Employer shall continue to enjoy his right under Clause 7(1)b and to copy and use all drawings specifications and other documents made by or on behalf of the Contractor.

The Employer may at any time sell any of the said Contractor's Equipment Temporary Works and unused goods and materials and apply the proceeds of sale in or towards the satisfaction of any sums due or which may become due to him from the Contractor under the Contract.

Assignment to Employer

(3) Where the Employer has entered upon the Works and any other parts of the Site as hereinbefore provided the Contractor shall if so instructed by the Engineer in writing within 7 days of such entry assign to the Employer the benefit of any agreement which the Contractor may have entered into for the supply of any goods or materials and/or for the carrying out of any work for the purposes of the Contract.

Valuation at date of termination

(4) As soon as may be practicable after any such entry and expulsion by the Employer the Engineer shall fix and determine as at the time of such entry and expulsion

(a) the amount (if any) which has been reasonably earned by or would reasonably accrue to the Contractor in respect of work actually done by him under the Contract and

(b) the value of any unused or partially used goods and materials which are under the control of the Employer

and shall certify accordingly.

The said determination may be carried out *ex parte* or by or after reference to the parties or after such investigation or enquiry as the Engineer may think fit to make or institute.

Payment after termination

(5) (a) If the Employer enters and expels the Contractor under this Clause he shall not be liable to pay the Contractor any money under the Contract (whether in respect of amounts certified by the Engineer or otherwise) unless or until the Engineer certifies that an amount is due to the Contractor under sub-clause (b).

(b) The Engineer shall certify the difference between

(i) such sum as would have been due to the Contractor if he had completed the Works together with any proceeds of sale under sub-clause 2 of this Clause and

(ii) the costs of completing the Works (whether or not the Works are completed under a separate contract) damages for delay (if any) and all other expenses properly incurred by the Employer.

(c) Such difference as is certified by the Engineer in sub-clause (b) shall be a debt due to the Employer or Contractor as the case may be.

(d) If the Engineer is satisfied at any time prior to the completion of the Works that such sum as calculated under sub-clause (b)(ii) exceeds such sum as calculated under sub-clause (b)(i) he may issue an interim certificate to that effect notwithstanding that the Works have not been completed and such interim certificate shall be considered a debt due from the Contractor to the Employer.

(e) Every certificate issued by the Engineer pursuant to this Clause shall be sent to the Employer and at the same time copied to the Contractor with such detailed explanation as may be necessary.

AVOIDANCE AND SETTLEMENT OF DISPUTES

Avoidance of disputes

66 (1) In order to overcome where possible the causes of disputes and in those cases where disputes are likely still to arise to facilitate their clear definition and early resolution (whether by agreement or otherwise) the following procedure shall apply for the avoidance and settlement of disputes.

Matters of dissatisfaction

(2) If at any time

(a) the Contractor is dissatisfied with any act or instruction of the Engineer's Representative or any other person responsible to the Engineer or

(b) the Employer or the Contractor is dissatisfied with any decision opinion instruction direction certificate or valuation of the Engineer or with any other matter arising under or in connection with the Contract or the carrying out of the Works

the matter of dissatisfaction shall be referred to the Engineer who shall notify his written decision to the Employer and the Contractor within one month of the reference to him.

Disputes

(3) The Employer and the Contractor agree that no matter shall constitute nor be said to give rise to a dispute unless and until in respect of that matter

(a) the time for the giving of a decision by the Engineer on a matter of dissatisfaction under Clause 66(2) has expired or the decision given is unacceptable or has not been implemented and in consequence the Employer or the Contractor has served on the other and on the Engineer a notice in writing (hereinafter called the Notice of Dispute) or

(b) an adjudicator has given a decision on a dispute under Clause 66(6) and the Employer or the Contractor is not giving effect to the decision and in consequence the other has served on him and the Engineer a Notice of Dispute

and the dispute shall be that stated in the Notice of Dispute. For the purposes of all matters arising under or in connection with the Contract or the carrying out of the Works the word "dispute" shall be construed accordingly and shall include any difference.

(4) (a) Notwithstanding the existence of a dispute following the service of a Notice under Clause 66(3) and unless the Contract has already been determined or abandoned the Employer and the Contractor shall continue to perform their obligations.

(b) The Employer and the Contractor shall give effect forthwith to every decision of

(i) the Engineer on a matter of dissatisfaction given under Clause 66(2) and

(ii) the adjudicator on a dispute given under Clause 66(6)

unless and until that decision is revised by agreement of the Employer and Contractor or pursuant to Clause 66.

Conciliation

(5) (a) The Employer or the Contractor may at any time before service of a Notice to Refer to arbitration under Clause 66(9) by notice in writing seek the agreement of the other for the dispute to be considered under "The Institution of Civil Engineers' Conciliation Procedure 1999" or any amendment or modification thereof being in force at the date of such notice.

(b) If the other party agrees to this procedure any recommendation of the conciliator shall be deemed to have been accepted as finally determining the dispute by agreement so that the matter is no longer in dispute unless a Notice of Adjudication under Clause 66(6) or a Notice to Refer to arbitration under Clause 66(9) has been served in respect of that dispute not later than one month after receipt of the recommendation by the dissenting party.

Adjudication

(6) (a) The Employer and the Contractor each has the right to refer a dispute as to a matter under the Contract for adjudication and either party may give notice in writing (hereinafter called the Notice of Adjudication) to the other at any time of his intention so to do. The adjudication shall be conducted under "The Institution of Civil Engineers' Adjudication Procedure 1997" or any amendment or modification thereof being in force at the time of the said Notice.

(b) Unless the adjudicator has already been appointed he is to be appointed by a timetable with the object of securing his appointment and referral of the dispute to him within 7 days of such notice.

(c) The adjudicator shall reach a decision within 28 days of referral or such longer period as is agreed by the parties after the dispute has been referred.

(d) The adjudicator may extend the period of 28 days by up to 14 days with the consent of the party by whom the dispute was referred.

(e) The adjudicator shall act impartially.

(f) The adjudicator may take the initiative in ascertaining the facts and the law.

(7) The decision of the adjudicator shall be binding until the dispute is finally determined by legal proceedings or by arbitration (if the contract provides for arbitration or the parties otherwise agree to arbitration) or by agreement.

(8) The adjudicator is not liable for anything done or omitted in the discharge or purported discharge of his functions as adjudicator unless the act or omission is in bad faith and any employee or agent of the adjudicator is similarly not liable.

Arbitration

(9) (a) All disputes arising under or in connection with the Contract or the carrying out of the Works other than failure to give effect to a decision of an adjudicator shall be finally determined by reference to arbitration. The party seeking arbitration shall serve on the other party a notice in writing (called the Notice to Refer) to refer the dispute to arbitration.

(b) Where an adjudicator has given a decision under Clause 66(6) in respect of the particular dispute the Notice to Refer must be served within three months of the giving of the decision otherwise it shall be final as well as binding.

Appointment of arbitrator

(10) (a) The arbitrator shall be a person appointed by agreement of the parties.

President or Vice-President to act

(b) If the parties fail to appoint an arbitrator within one month of either party serving on the other party a notice in writing (hereinafter called the Notice to Concur) to concur in the appointment of an arbitrator the dispute shall be referred to a person to be appointed on the application of either party by the President for the time being of the Institution of Civil Engineers.

(c) If an arbitrator declines the appointment or after appointment is removed by order of a competent court or is incapable of acting or dies and the parties do not within one month of the vacancy arising fill the vacancy then either party may apply to the President for the time being of the Institution of Civil Engineers to appoint another arbitrator to fill the vacancy.

(d) In any case where the President for the time being of the Institution of Civil Engineers is not able to exercise the functions conferred on him by this Clause the said functions shall be exercised on his behalf by a Vice-President for the time being of the said Institution.

Arbitration — procedure and powers

(11) (a) Any reference to arbitration under this Clause shall be deemed to be a submission to arbitration within the meaning of the Arbitration Act 1996 or any statutory re-enactment or amendment thereof for the time being in force. The reference shall be conducted in accordance with the procedure set out in the Appendix to the Form of Tender or any amendment or modification thereof being in force at the time of the appointment of the arbitrator. Such arbitrator shall have full power to open up review and revise any decision opinion instruction direction certificate or valuation of the Engineer or an adjudicator.

(b) Neither party shall be limited in the arbitration to the evidence or arguments put to the Engineer or to any adjudicator pursuant to Clause 66(2) or 66(6) respectively.

(c) The award of the arbitrator shall be binding on all parties.

(d) Unless the parties otherwise agree in writing any reference to arbitration may proceed notwithstanding that the Works are not then complete or alleged to be complete.

Witnesses

(12) (a) No decision opinion instruction direction certificate or valuation given by the Engineer shall disqualify him from being called as a witness and giving evidence before a conciliator adjudicator or arbitrator on any matter whatsoever relevant to the dispute.

(b) All matters and information placed before a conciliator pursuant to a reference under sub-clause (5) of this Clause shall be deemed to be submitted to him without prejudice and the conciliator shall not be called as witness by the parties or anyone claiming through them in connection with any adjudication arbitration or other legal proceedings arising out of or connected with any matter so referred to him.

APPLICATION TO SCOTLAND AND NORTHERN IRELAND

Application to Scotland

67 (1) If the Works are situated in Scotland (and unless the Contract otherwise provides) the Contract shall in all respects be construed and operate as a Scottish contract and shall be interpreted in accordance with Scots Law and the provisions of sub-clause (2) of this Clause shall apply.

(2) In the application of these Conditions and in particular Clause 66 thereof

(a) any reference to arbitration under these Conditions shall be conducted in accordance with the law of Scotland "The Scottish Arbitration Code 1999" prepared by the Scottish Council for International Arbitration the Chartered Institute of Arbitrators (Scottish Branch) and the Scottish Building Contract Committee together with the ICE Appendix (2001) thereto or any amendment to or modification of the Appendix being in force at the time of appointment of the arbitrator. Such arbitrator shall have full power to open up review and revise any decision opinion instruction direction certificate or valuation of the Engineer or an adjudicator

(b) for any reference to the "Notice to Refer" there shall be substituted reference to the "Notice of Arbitration"

(c) the existing Clause 66(10) shall be deleted and replaced with new Clause 66(10) below

" **Appointment of arbitrator**

66(10) (a) The arbitral tribunal shall be appointed by agreement of the parties.

President or Vice-President to act

(b) Failing agreement of the parties as aforesaid at sub-clause (a) above the following shall apply.

(i) Reference at Articles 3.5 3.6 3.7 3.8 and 6.2 of the Code to the Chairman of the Institute of Arbitrators (Scottish Branch) and to the Chairman of the Scottish Council for International Arbitration shall be deemed to be a reference to the President of the Institution of Civil Engineers as defined at (ii) below.

(ii) "President" means the President for the time being of the Institution of Civil Engineers or any Vice President acting on his behalf or such other person as may have been nominated in the arbitration agreement to appoint the arbitrator in default of agreement between the parties. "

(d) the existing Clause 66(11) shall be deleted and replaced with new Clause 66(11) below

" Arbitration – procedure and powers **66(11)** (a) Neither party shall be limited in the arbitration to the evidence or arguments put to the Engineer or to any adjudicator pursuant to Clause 66 (2) or 66(6) respectively.

(b) Unless the parties otherwise agree in writing any reference to arbitration may proceed notwithstanding that the Works are not then complete or alleged to be complete."

(e) notwithstanding any of the other provisions of these Conditions or of the Code (including in particular Articles 22.7 and 22.8) nothing therein shall exclude or be construed as excluding recourse to the Court of Session under Section 3 of the Administration of Justice of (Scotland) Act 1972

(f) where the Employer or the Contractor wishes to register the award of an arbitrator or the decision of an adjudicator in the Books of Council and Session for preservation and execution the other party shall on being requested to do so forthwith consent to such registration by subscribing the said decision before a witness

(g) delete existing Appendix to the Form of Tender Part 1 Item 22 and

(h) the application of Articles 1.6 1.8 1.9 and 1.10 of the Scottish Arbitration Code 1999 shall be subject to the provisions of Clause 66(2) of the Contract unless the Arbitrator otherwise directs. "

Application to Northern Ireland

(3) If the Works are situated in Northern Ireland the Contract shall in all respects be construed and operate as a Northern Irish contract and shall be interpreted in accordance with the law of Northern Ireland.

Application elsewhere

(4) If the Works are situated in a country or jurisdiction other than England and Wales Scotland or Northern Ireland the Contract and the provisions for disputes settlement shall in all respects be construed and operate and be interpreted in accordance with the law of that country or jurisdiction.

NOTICES

Service of notices on Contractor

68 (1) Any notice to be given to the Contractor under the terms of the Contract shall be served in writing at the Contractor's principal place of business (or in the event of the Contractor being a Company to or at its registered office).

Service of notices on Employer

(2) Any notice to be given to the Employer under the terms of the Contract shall be served in writing at the Employer's last known address (or in the event of the Employer being a Company to or at its registered office).

(3) Any notice to be given by or to the Engineer under the terms of the Contract shall be delivered as the Engineer may direct.

TAX MATTERS

Labour-tax and landfill tax fluctuations

69 (1) The rates and prices contained in the Bill of Quantities shall be deemed to take account only of the levels and incidence in force at the date for return of tenders of

(a) the taxes levies contributions premiums or refunds (including national insurance contributions but excluding income tax and any levy payable under the Industrial Training Act 1982 or any statutory re-enactment or amendment thereof for the time being in force) which are by law payable by or to the Contractor and his sub-contractors in respect of their workpeople engaged on the Contract and

(b) any landfill tax payable by the Contractor or his sub-contractors pursuant to the Finance Act 1996 (Sections 39-71 and Schedule 5) and the Landfill Tax Regulations 1996 or any statutory re-enactment or amendment thereof for the time being in force

(c) any aggregate levy payable by the Contractor or his sub-contractors pursuant to the Finance Act 2001 or any statutory re-enactment or amendment thereof for the time being in force

and shall not take account of any level or incidence of the aforesaid matters foreseeable or known to take effect at some later date.

(2) If after the date for return of tenders there shall occur any change in the level and/or incidence of any such taxes levies contributions premiums or refunds the Contractor shall so inform the Engineer and the net increase or decrease shall be taken into account in arriving at the Contract Price. The Contractor shall supply the information necessary to support any consequent adjustment to the Contract Price. All certificates for payment issued after submission of such information shall take due account of the additions or deductions to which such information relates.

Value Added Tax

70 (1) The Contractor shall be deemed not to have allowed in his tender for the tax payable by him as a taxable person to the Commissioners of Customs and Excise being tax chargeable on any taxable supplies to the Employer which are to be made under the Contract.

Engineer's certificates net of Value Added Tax

(2) All certificates issued by the Engineer under Clause 60 shall be net of Value Added Tax.

In addition to the payments due under such certificates the Employer shall separately identify and pay to the Contractor any Value Added Tax properly chargeable by the Commissioners of Customs and Excise on the supply to the Employer of any goods and/or services by the Contractor under the Contract.

Disputes

(3) If any dispute difference or question arises between either the Employer or the Contractor and the Commissioners of Customs and Excise in relation to any tax chargeable or alleged to be chargeable in connection with the Contract or the Works each shall render to the other such support and assistance as may be necessary to resolve the dispute difference or question.

Clause 66 not applicable

(4) Clause 66 shall not apply to any dispute difference or question arising under this Clause.

THE CONSTRUCTION (DESIGN AND MANAGEMENT) REGULATIONS 1994

CDM Regulations 1994 **71** (1) In this clause

(a) "the Regulations" means the Construction (Design and Management) Regulations 1994 or any statutory re-enactment or amendment thereof for the time being in force

(b) "Planning Supervisor" and "Principal Contractor" mean the persons so described in regulation 2(1) of the Regulations

(c) "Health and Safety Plan" means the plan prepared by virtue of regulation 15 of the Regulations.

(2) Where and to the extent that the Regulations apply to the Works and

(a) the Engineer is appointed Planning Supervisor and/or

(b) the Contractor is appointed Principal Contractor

then in taking any action as such they shall state in writing that the action is being taken under the Regulations.

(3) (a) Any action under the Regulations taken by either the Planning Supervisor or the Principal Contractor and in particular any alteration or amendment to the Health and Safety Plan shall be deemed to be an Engineer's instruction pursuant to Clause 13. Provided that the Contractor shall in no event be entitled to any additional payment and/or extension of time in respect of any such action to the extent that it results from any action lack of action or default on the part of the Contractor.

(b) If any such action of either the Planning Supervisor or the Principal Contractor could not in the Contractor's opinion reasonably have been foreseen by an experienced contractor the Contractor shall as early as practicable give written notice thereof to the Engineer.

SPECIAL CONDITIONS

Special conditions **72** The following special conditions form part of the Conditions of Contract.

(Note. Any special conditions including Contract Price Fluctuation which it is desired to incorporate in the Conditions of Contract should be numbered consecutively with the foregoing conditions of contract).

SHORT DESCRIPTION OF WORKS

All Permanent and Temporary Works in connection with* ....

....

....

Form of Tender

(NOTE: The Appendix forms part of the Form of Tender)

To ....

....

....

OFFER

GENTLEMEN,

Having examined the Drawings, Conditions of Contract, Specification and Bill of Quantities for the construction of the above-mentioned Works (and the matters set out in the Appendix hereto) we offer to construct and complete the whole of the said Works in conformity with the said Drawings, Conditions of Contract, Specification and Bill of Quantities for such sum as may be ascertained in accordance with the said Conditions of Contract.

We undertake to complete and deliver the whole of the Permanent Works comprised in the Contract within the time stated in the Appendix hereto.

If our tender is accepted we will, if required, provide security for the due performance of the Contract as stipulated in the Conditions of Contract and the Appendix hereto.

Unless and until a formal Agreement is prepared and executed this tender together with your written acceptance thereof, shall constitute a binding Contract between us.

We understand that you are not bound to accept the lowest or any tender you may receive.

We are, Gentlemen.

Yours faithfully,

Signature ....

Address ....

....

Date ....

* Complete as appropriate

FORM OF TENDER (APPENDIX)

(NOTE: Relevant Clause numbers are shown in brackets)

Appendix - Part 1 (to be completed prior to the invitation to tender)

1 Name of the Employer (Clause 1(1)(a)) ....

Address ....

2 Name of the Engineer (Clause 1(1)(c)) ....

Address...

3 Defects Correction Period (Clause 1(1)(s)) ....weeks

4 Parts or Sections of the Works which shall not be sub-contracted without the Engineer's prior written approval (Clause 4(2))

....

....

....

5 Number and type of copies of Drawings to be provided (Clause 6(1)(b)) ....

....

6 Form of Agreement (Clause 9) Required/Not required

If required Signed/Sealed

7 Performance Bond (Clause 10(1)) Required/Not required

Amount of Bond (if required) to be ....% of Tender Total

8 Minimum amount of third party insurance (persons and property) (Clause 23(3)) £ .... for each and every occurrence

9 Works Commencement Date (if known) (Clause 41(1)(a)) ....

10 Time for Completion (Clause 43) [a]

EITHER for the whole of the Works ....weeks

OR for Sections of the Works (Clause 1(1)(u)) [b]

Section A ....weeks

Section B ....weeks

Section C ....weeks

Section D ....weeks

the Remainder of the Works ....weeks

11 Liquidated damages for delay (Clause 47)

	per day/week	limit of liability [c]
EITHER for the whole of the Works	£............................	£...............................
OR for Section A (as above)	£............................	£...............................
Section B (as above)	£............................	£...............................
Section C (as above)	£............................	£...............................
Section D (as above)	£............................	£...............................
the Remainder of the Works (as above)	£............................	£...............................

12 Vesting of materials not on Site (Clauses 54(4) and 60(1)(c)) (if required by the Employer) [d]

1.. 4...

2.. 5..

3.. 6..

13 Method of measurement adopted in preparation of Bills of Quantities (Clause 57) [e]...............................

....

14 Percentage of the value of goods and materials to be included

in Interim Certificates (Clause 60(2)(b)) ....%

15 Minimum amount of Interim Certificates (Clause 60(3)) £..............

16 Rate of retention (recommended not to exceed 5%) (Clause 60(5)) ....%

17 Limit of retention (% of Tender Total) (Clause 60(5)) (Recommended not to exceed 3%) ....%

18 Bank whose Base Lending Rate is to be used (Clause 60(7)) ....

19 Requirement for prior approval by the Employer before the Engineer can act.
DETAILS TO BE GIVEN AND CLAUSE NUMBER STATED (Clause 2(1)(b)) [f]

....

....

....

20 Name of the Planning Supervisor (Clause 71(1)(b))

....

Address ....

21 Name of the Principal Contractor (if appointed) (Clause 71(1)(b))

....

Address ....

22 The arbitration procedure to be used is (Clause 66(11)(a)

(a) The Institution of Civil Engineers' Arbitration Procedure 1997 [g] or

(b) The Construction Industry Model Arbitration Rules [g]

If no deletion is made, the Institution of Civil Engineers' Arbitration Procedure 1997 shall be used.

[a] If not stated is to be completed by Contractor in Part 2 of the Appendix.

[b] To be completed if required, with brief description. Where Sectional completion applies the item for "the Remainder of the Works" must be used to cover the balance of the Works if the Sections described do not in total comprise the whole of the Works.

[c] Delete where not required.

[d] (If used) Materials to which the Clauses apply must be listed in Part I (Employer's option) or Part 2 (Contractor's option)

[e] Insert here any amendment or modification adopted if different from that stated in Clause 57.

[f] If there is any requirement that the Engineer has to obtain prior approval from the Employer before he can act full particulars of such requirements must be set out above.

[g] Delete one as appropriate.

Appendix - Part 2

(To be completed by Contractor)

1 Insurance Policy Excesses (Clause 25(2))

Insurance of the Works (Clause 21(1)) £.....................

Third party (property damage) (Clause 23(1)) £.....................

2 Time for Completion (Clause 43) (if not completed in Part I of the Appendix)

EITHER for the whole of the Works ....weeks

OR for Sections of the Works (Clause l(1)(u)) (as detailed in Part 1 of the Appendix)

Section A ....weeks

Section B ....weeks

Section C ....weeks

Section D ....weeks

the Remainder of the Works ....weeks

3 Vesting of materials not on site (Clauses 54(4) and 60(1)(c)) (at the option of the Contractor — see [d] in Part 1)

1 .... 4 ....

2 .... 5 ....

3 .... 6 ....

4 Percentage(s) for adjustment of PC sums (Clauses 59(2)(c) and 59(5)(c)) (with details if required)

....

....

Form of Agreement

THIS AGREEMENT made the .... day of .... 19 ....

BETWEEN ....

of...

in the County of ....(hereinafter called "the Employer")

and ....

of ....

in the County of...(hereinafter called "the Contractor").

WHEREAS the Employer is desirous that certain Works should be constructed, namely the Permanent and Temporary Works in connection with..

....

and has accepted a Tender by the Contractor for the construction and completion of such Works.

NOW THIS AGREEMENT WITNESSETH as follows

1. In this Agreement words and expressions shall have the same meanings as are respectively assigned to them in the Conditions of Contract hereinafter referred to.

2. The following documents shall be deemed to form and be read and construed as part of this Agreement, namely

(a) the said Tender and the written acceptance thereof
(b) the Drawings
(c) the Conditions of Contract
(d) the Specification
(e) the priced Bill of Quantities.

3. In consideration of the payments to be made by the Employer to the Contractor as hereinafter mentioned the Contractor hereby covenants with the Employer to construct and complete the Works in conformity in all respects with the provisions of the Contract.

4. The Employer hereby covenants to pay to the Contractor in consideration of the construction and completion of the Works the Contract Price at the times and in the manner prescribed by the Contract.

IN WITNESS whereof the parties hereto have caused this Agreement to be executed the day and year first above written.

SIGNED on behalf of the said ....Ltd/plc (the Employer)

Signature .... Signature ....

Position .... Position ....

In the presence of .... In the presence of ....

SIGNED on behalf of the said ....Ltd/plc (the Contractor)

Signature .... Signature ....

Position .... Position ....

In the presence of .... In the presence of ....

or

SIGNED [and SEALED*] AS A DEED by the said ....

....Ltd/plc (the Employer)

In the presence of ....

or

SIGNED [and SEALED*] AS A DEED by the said ....

....Ltd/plc (the Contractor)

In the presence of ....

* Delete as appropriate

ICE FORM OF DEFAULT BOND

Date

Parties

[Names addresses and Company Numbers if applicable]

SURETY (1)

CONTRACTOR (2)

EMPLOYER (3)

Background

(A) By a Contract defined in the Schedule hereto the Contractor has agreed with the Employer to construct and complete the Works.

(B) The Surety has agreed to provide this Bond in favour of the Employer in order to guarantee the performance by the Contractor of his obligations under the Contract.

Surety's obligation **1**

(1) If the Contractor fails to pay the Excess Sum within 28 days of receipt by the Contractor of a copy of a certificate issued by the Engineer under clause 65(5) of the Contract the Surety hereby guarantees to the Employer that the Surety shall subject to the terms and conditions of this Bond pay the Excess Sum in accordance with Clause 1(3) up to the Bond Amount.

(2) It shall be a condition precedent to payment by the Surety that the Employer serve on the Surety a copy of the certificate issued by the Engineer under clause 65(5) of the Contract as served on the Contractor and certified by the Engineer as being a true copy of such certificate.

(3) Subject to Clause 1(4) payment by the Surety shall be made not later than 14 days after the later of

(a) the expiry of the 28 day period referred to in Clause 1(1) (save in respect of any payment made by the Contractor within that time) and

(b) service on the Surety of the copy certificate referred to in Clause 1(2).

(4) If the Surety objects to the contents of or entitlement to issue a certificate under clause 65(5) of the Contract in respect of which the Employer seeks payment from him the Surety shall have the right to refer the matter to adjudication in accordance with the adjudication provisions contained in sub-clauses 66(6) to 66(8) of the Contract as if the Surety were a party to the Contract in place of the Contractor.

(5) Any adjudication under Clause 1(4) shall be commenced by the Surety within 14 days of receipt by the Surety of the documents referred to in Clause 1(2) and the Surety shall have no right to refer the matter to adjudication after that time.

(6) If the content of or entitlement to issue the Certificate under clause 65(5) of the Contract in respect of which payment is sought by the Employer is or has been the subject of an adjudication between the Employer and Contractor under the Contract (in respect of which both parties have made submissions to the adjudicator) the Surety agrees to be bound by the result of such adjudication and shall have no right to refer the matter to adjudication under Clause 1(4).

(7) In the case of an adjudication under Clause 1(4) payment by the Surety shall be made within 7 days of the decision in such adjudication.

Surety's rights 2 (1) The Surety shall be entitled to receive copies of any notice given by the Employer under clause 65(1) of the Contract (with any Engineer's certificate referred to) within 7 days of such notice or certificate being served on the Contractor.

(2) The Surety shall be entitled at any time within 7 days of receipt by the Surety of the copy certificate referred to in Clause 1(2)

(a) to request the Employer to provide the Surety with such further information and documentation as the Surety may reasonably require to verify the Excess Sum (including information or documentation held by the Engineer) and/or

(b) to request to inspect the Site and the Works upon reasonable notice (the Employer may require that a representative of the Employer accompanies the Surety during such inspections).

Accounting 3 (1) If the Excess Sum is subsequently determined by reason of a subsequent certificate issued by the Engineer under clause 65(5) of the Contract or by adjudication arbitration litigation or agreement between the Surety and the Employer to be less than the amount paid by the Surety the difference (if the Excess Sum has already been paid by the Surety) shall be repaid by the Employer to the Surety with Interest within 14 days (or such other period as the adjudicator arbitrator or Court may direct) after the date of such determination or agreement.

(2) If the Excess Sum is subsequently determined or agreed to be greater than the amount already paid by the Surety any difference (up to the Bond Amount) shall be paid by the Surety to the Employer within 14 days (or such other period as the adjudicator arbitrator or Court may direct) after the date of such determination or agreement.

Interest 4 Subject to the amount payable by the Surety being varied in accordance with Clause 3 above in the event that any amount payable by either the Surety or the Employer under this Bond is not made by the date determined by Clause 1(3) or in accordance with Clause 3 (the Due Date) then the payer shall pay Interest on the sum from the Due Date until the date of payment.

Expiry	**5**	Save in respect of any failure to pay the Excess Sum in respect of which a claim in writing has been received beforehand from the Employer by the Surety this Bond shall expire on the earlier of the date stated in a certificate of substantial completion issued by the Engineer and the Final Expiry Date.
Forbearance	**6**	The Surety shall not be discharged or released by any alteration variation or waiver of any of the terms and conditions and provisions of the Contract or in any extent or nature of the Works and no allowance of time by the Employer under or in connection with the Contract or the Works shall in any way release reduce or affect the liability of the Surety under this Bond.
Governing law	**7**	This Bond shall be governed and construed in accordance with the laws of the country named in the Schedule ("the Country") and the courts of the Country shall have exclusive jurisdiction.
Assignment	**8**	This Bond may only be assigned by the Employer with the prior consent of the Surety and the Contractor which consent shall not be unreasonably withheld. In the event of any such assignment the Employer and assignee shall remain jointly and severally liable for any repayment due to the Surety under Clause 3(1). Notice of any assignment shall be given to the Surety as soon as practicable.

EXECUTED AS A DEED

on behalf of the Surety	(1)	director	________________________
	(2)	director/secretary	________________________
on behalf of the Contractor	(1)	director	________________________
	(2)	director/secretary	________________________
on behalf of the Employer	(1)	director	________________________
	(2)	director/secretary	________________________

SCHEDULE

Address for Service	**Contractor:**
	Tel: Fax:
	Employer:
	Tel: Fax:
	Surety:
	Tel: Fax:
"Bond Amount"	means the sum of £[ ] ([ ] pounds) being the maximum aggregate liability of the Surety under this Bond.
"Contract"	means the contract [made between the Employer and the Contractor dated the [ ] day of [ ] [ ]] / [to be entered into between the Employer and the Contractor] incorporating the ICE Conditions of Contract 7th Edition.
"Excess Sum"	an amount certified as due to the Employer under Clause 65(5) of the Contract.
"Engineer" "Works" and "Site"	have the same meaning as in the Contract.
"Final Expiry Date"	means the [ ] day of [ ] [ ].
"Interest"	means the rate of interest specified in the Contract.
"The Country"	means [England and Wales] [Scotland] [Northern Ireland]

(This Bond has been drafted in collaboration with S. J. BERWIN & Co)

The Institution of Civil Engineers | The Association of Consulting Engineers | The Civil Engineering Contractors Association

This clause has been prepared by The Institution of Civil Engineers, The Association of Consulting Engineers and The Civil Engineering Contractors Association, in consultation with the Government in its revised form, for use in appropriate cases as a Special Condition of the Conditions of Contract for use in connection with the ICE Conditions of Contract SEVENTH EDITION dated September 1999.

CONTRACT PRICE FLUCTUATIONS CIVIL ENGINEERING WORK

(1) The amount payable by the Employer to the Contractor upon the issue by the Engineer of an interim certificate pursuant to Clause 60(2) or of the final certificate pursuant to Clause 60(4) (other than amounts due under this Clause) shall be increased or decreased in accordance with the provisions of this Clause if there shall be any changes in the following Index Figures compiled by the Department of the Environment Transport and the Regions and published by The Stationery Office in the Monthly Bulletin of Indices "1990 Series Civil Engineering Formula Indices" and "1990 Series Structural Steelwork Formula Indices"

(a) the Index of Labour and Supervision (Index 1)

(b) the Index of providing and maintaining Contractor's Equipment (Index 2 - Plant and Road Vehicles)

(c) the Indices of material prices applicable to those materials listed in sub-clause (4) of this Clause.

The net total of such increases and decreases shall be given effect to in determining the Contract Price.

(2) For the purpose of this Clause

(a) 'Final Index Figure' shall mean any Index Figure appropriate to sub-clause (1) of this Clause not qualified in the said bulletin as provisional

(b) 'Base Index Figure' shall mean the appropriate Final Index Figure applicable to the date 42 days prior to the date for the return of tenders

(c) 'Current Index Figure' shall mean the appropriate Final Index Figure to be applied in respect of any certificate issued or due to be issued by the Engineer pursuant to Clause 60 and shall be the appropriate Final Index Figure applicable to the date 42 days prior to

(i) the due date (or extended date) for completion or

(ii) the date certified pursuant to Clause 48 of completion of the whole of the Works or

(iii) the last day of the period to which the certificate relates

whichever is the earliest.

Provided that in respect of any work the value of which is included in any such certificate and which work forms part of a Section for which the due date (or extended date) for completion or the date certified pursuant to Clause 48 of completion of such Section precedes the last day of the period to which the certificate relates the Current Index Figure shall be the Final Index Figure applicable to the date 42 days prior to whichever of these dates is the earliest.

(d) The 'Effective Value' in respect of the whole or any Section of the Works shall be the difference between

(i) the amount which in the opinion of the Engineer is due to the Contractor under Clause 60(2) (before deducting retention) or the amount due to the Contractor under Clause 60(4) (but in each case before deducting sums previously paid on account) less any amounts for Dayworks Nominated Sub-contractors or any other items based on actual cost or current prices and any sums for increases or decreases in the Contract Price under this Clause and

(ii) the amount calculated in accordance with (i) above and included in the last preceding interim certificate issued by the Engineer in accordance with Clause 60.

Provided that in the case of the first certificate the Effective Value shall be the amount calculated in accordance with sub-paragraph (i) above.

(3) The increase or decrease in the amounts otherwise payable under Clause 60 pursuant to sub-clause (1) of this Clause shall be calculated by multiplying the Effective Value by a Price Fluctuation Factor which shall be the net sum of the products obtained by multiplying each of the proportions given in (a) (b) and (c) of sub-clause (4) of this Clause by a fraction the numerator of which is the relevant Current Index Figure minus the relevant Base Index Figure and the denominator of which is the relevant Base Index Figure.

(4) For the purpose of calculating the Price Fluctuation Factor the proportions referred to in sub-clause (3) of this Clause shall (irrespective of the actual constituents of the work) be as follows and the total of such proportions shall amount to unity

(a) 0.____ * in respect of labour and supervision costs subject to adjustment by reference to the Index referred to in sub-clause (1)(a) of this Clause. (Index 1)

(b) 0.____ * in respect of costs of provision and use of Contractor's Equipment which shall be subject to adjustment by reference to the Index referred to in sub-clause (1)(b) of this Clause. (Index 2)

(c) the following proportions in respect of the materials named which shall be subject to adjustment by reference to the relevant indices referred to in sub-clause (1)(c) of this Clause

0.____ *	in respect of Aggregates. (Index 3)
0.____ *	in respect of Bricks and Clay Products. (Index 4)
0.____ *	in respect of Cements. (Index 5)
0.____ *	in respect of Ready Mixed Concrete. (Index 6)
0.____ *	in respect of Cast and Spun Iron Products. (Index 7)
0.____ *	in respect of Plastics Products. (Index 8)
0.____ *	in respect of Coated Macadam and Bituminous Products. (Index 9)
0.____ *	in respect of Fuel for plant to which the DERV Fuel Index will be applied. (Index 10)
0.____ *	in respect of Fuel for plant to which the Gas Oil Fuel Index will be applied. (Index 11)
0.____ *	in respect of Timber. (Index 12)
0.____ *	in respect of Steel for Reinforcement. (Index 13)
0.____ *	in respect of Metal Sections. (Index 14)
0.____ *	in respect of Sheet Steel Piling. (Index 15)
0.____ *	in respect of Structural Steelwork Materials for Civil Engineering Works (Index S3)

(d) 0. 10 in respect of all other costs which shall not be subject to any adjustment

Total 1.00

(5) Provisional Index Figures in the Bulletin referred to in sub-clause (1) of this Clause may be used for the provisional adjustment of interim valuations but such adjustments shall be subsequently recalculated on the basis of the corresponding Final Index Figures.

(6) Clause 69 – Tax Fluctuations – shall not apply except to the extent that any matter dealt with therein is not covered by the Index of the Cost of Labour in Civil Engineering Construction.

* To be filled in by the Employer prior to inviting tenders.

The Institution of Civil Engineers | The Association of Consulting Engineers | The Civil Engineering Contractors Association

This clause has been prepared by The Institution of Civil Engineers, The Association of Consulting Engineers and The Civil Engineering Contractors Association, in consultation with the Government in its revised form, for use in appropriate cases as a Special Condition of the Conditions of Contract for use in connection with the ICE Conditions of Contract SEVENTH EDITION dated September 1999.

CONTRACT PRICE FLUCTUATIONS STRUCTURAL STEELWORK

(1) The amount payable by the Employer to the Contractor upon the issue by the Engineer of an interim certificate pursuant to Clause 60(2) or of the final certificate pursuant to Clause 60(4) (other than amounts due under this Clause) shall be increased or decreased in accordance with the provisions of this Clause if there shall be any changes in the following Index Figures compiled by the Department of the Environment Transport and the Regions and published by the Stationery Office in the Monthly Bulletin of Indices "1990 Series Structural Steelwork Formula Indices"

(a) the Index of Labour in fabrication of Structural Steelwork and erection of Structural Steelwork (Index S1)

(b) the Index for Structural Steelwork Materials (Civil Engineering Works) (Index S3).

The net total of such increases and decreases shall be given effect to in determining the Contract Price.

(2) For the purpose of this Clause

(a) 'Structural Steelwork' shall mean those items of work listed in sub-clause (6) of this Clause and shall include any variations as may be ordered under Clause 51 involving work of a description which in the opinion of the Engineer is similar to the description of the items so listed

(b) 'Final Index Figure' shall mean any Index Figure appropriate to sub-clause (1) of this Clause not qualified in the said bulletin as provisional

(c) 'Base Index Figure' shall mean the appropriate Final Index Figure applicable to the date 42 days prior to the date for the return of tenders

(d) 'Current Index Figure' shall mean the appropriate Final Index Figure to be applied in respect of any certificate issued or due to be issued by the Engineer pursuant to Clause 60 being such figure applicable

(i) in respect of labour employed in fabrication - to a date 56 days prior to the last day of the period to which the certificate relates or

(ii) in respect of labour employed in erection - to a date 14 days prior to the last day of the period to which the certificate relates or

(iii) in respect of materials specifically purchased for inclusion in the Works - to the date of delivery to the fabricator's premises (of which date the Contractor shall produce such evidence relating to gross tonnages delivered as the Engineer may reasonably require) or

(iv) in respect of materials (if any) not specifically purchased for inclusion in the Works – to the date of the last of the deliveries referred to in sub-paragraph (iii) of this paragraph

as the case may be.

Provided always that should the due date (or extended date) for completion or the date certified pursuant to Clause 48 of completion of the whole of the Works precede any of the aforesaid then such due date or extended date or certified date whichever is earliest shall be substituted for those aforesaid.

Provided further that if in respect of any work which forms part of a Section and whose value is included in any certificate the due date (or extended date) for completion of that Section or the date certified pursuant to Clause 48 of completion of that Section precede any of the dates aforesaid in sub-paragraphs (i) to (iv) above then such due date or extended date or certified date whichever is the earliest in respect of that Section shall be substituted for those aforesaid.

(e) The 'Effective Value' in respect of the whole or any Section of the Works shall be the difference between

(i) the amount which in the opinion of the Engineer is due to the Contractor under Clause 60(2) (before deducting retention) or the amount due to the Contractor under Clause 60(4) (but in each case before deducting sums previously paid on account) less any amounts for Dayworks Nominated Sub-contractors or any other items based on actual cost or current prices and any sums for increases or decreases in the Contract Price under this Clause and

(ii) the amount calculated in accordance with (i) above and included in the last preceding interim certificate issued by the Engineer in accordance with Clause 60.

Provided that in the case of the first certificate the Effective Value shall be the amount calculated in accordance with sub-paragraph (i) above.

(3) The Effective Value shall be apportioned between labour and materials in the following manner that is to say

(a) labour

(i) employed in erection - by multiplying the total tonnage erected during the period to which the certificate relates by the average cost per tonne entered at (b) of sub-clause (5) of this Clause

(ii) employed in fabrication and delivery - by deducting the summation of the values calculated in respect of materials and in respect of labour employed in erection from the Effective Value

(b) materials

by multiplying the total tonnage of steel delivered to the Site for inclusion in the Works during the period to which the certificate relates by the average price per tonne entered at (a) of sub-clause (5) of this Clause.

(4) (a) The increase or decrease in the amounts otherwise payable under Clause 60 pursuant to sub-clause (1) of this Clause shall be calculated by multiplying each portion of the Effective Value by a fraction the numerator of which is the product of 0.90 and the difference between the relevant Current Index Figure and the relevant Base Index Figure and the denominator of which is the relevant Base Index Figure.

(b) The relevant indices to be used in connection with this sub-clause are

(i) for labour in fabrication and erection-the Index referred to in sub-clause (1)(a) of this Clause (Index S1)

(ii) for materials – the Index referred to in sub-clause (1)(b) of this Clause (Index S3).

(5) For the purpose of the apportionment in sub-clause (3) of this Clause the full average costs per tonne (inclusive of all associated labour plant power maintenance overheads and profit) to be used are

(a) Materials delivered to fabricators premises £ per tonne *

(b) Erection £ per tonne *

(c) Subject to sub-paragraphs (i) (ii) and (iii) of this paragraph the relevant figures to be used in connection with this sub-clause shall be in accordance with sub-clause (2) of this Clause.

Provided that in respect of materials

(i) the Current Index Figure subsequent to the first established Current Index Figure pursuant to sub-clause (2) (d) (iii) of this Clause shall not be used until a tonnage of steel greater than the tonnage to which the Current Index Figure first established applies has been delivered to the Site for inclusion in the Works and

(ii) for the purpose of establishing the appropriate subsequent Current Index Figures to apply to all later deliveries of steel to the Site for inclusion in the Works the provisions of sub-paragraph (i) of this paragraph shall apply mutatis mutandis and

(iii) the Current Index Figure referred to in sub-clause (2) (d) (iv) of this Clause shall not be used until the total tonnage of steel delivered to the Site for inclusion in the Works exceeds the total tonnage of steel specifically purchased for inclusion in the Works and delivered to the fabricator's premises.

(6) For the purposes of this Clause the expression 'Structural Steelwork' shall comprise only those items listed hereunder †

Bill No †	Page No †	Item No †

(7) Provisional Index Figures in the Bulletin referred to in sub-clause (1) of this Clause may be used for the provisional adjustment of interim valuations but such adjustments shall be subsequently recalculated on the basis of the corresponding Final Index Figures.

(8) Clause 69 – Tax Fluctuations – shall not apply except to the extent that any matter dealt with therein is not covered by the Index of the Cost of Labour in fabrication of steelwork and erection of steelwork.

* To be filled in by the Contractor at the time of tendering.
† To be filled in by the Employer prior to inviting tenders.

The Institution of Civil Engineers | The Association of Consulting Engineers | The Civil Engineering Contractors Association

This clause has been prepared by The Institution of Civil Engineers, The Association of Consulting Engineers and The Civil Engineering Contractors Association, in consultation with the Government in its revised form, for use in appropriate cases as a Special Condition of the Conditions of Contract for use in connection with the ICE Conditions of Contract SEVENTH EDITION dated September 1999.

CONTRACT PRICE FLUCTUATIONS CIVIL ENGINEERING WORK AND STRUCTURAL STEELWORK

(1) This Clause shall apply only to those contracts which incorporate the Contract Price Fluctuation Clause Revised February 1999 for Civil Engineering work and the Structural Steelwork Clause Revised February 1999 (referred to in this Clause as the CEW Clause and the SS Clause respectively).

(2) For the purposes of this Clause

(a) 'Civil Engineering work' shall mean all Works with the exception of Structural Steelwork and

(b) 'Structural Steelwork' shall mean the work defined in sub-clause (2)(a) of the SS Clause.

(3) The Effective Value (as defined both in the CEW Clause and in the SS Clause) shall be sub-divided to show the amounts included in respect of the Civil Engineering work and the Structural Steelwork. The amount in respect of the former shall then be treated as if it were the Effective Value as defined in the CEW Clause and adjusted in accordance with the provisions of that Clause. The amount in respect of the latter shall then be treated as if it were the Effective Value as defined in the SS Clause and adjusted in accordance with that Clause.